BUDDHISM AND THE COSMOS

Daisaku Ikeda in conversation with
Masayoshi Kiguchi and Eiichi Shimura

BQ
4570
.C6
I43x
West

Macdonald

A Macdonald Book

Copyright © Daisaku Ikeda 1985

First published in Great Britain in 1985 by
Macdonald & Co (Publishers) Ltd
London & Sydney
Reprinted 1986

All rights reserved

No part of this publication may be reproduced, stored in a retrieval system, or transmitted, in any form or by any means without the prior permission in writing of the publisher, nor be otherwise circulated in any form of binding or cover other than that in which it is published and without a similar condition including this condition being imposed on the subsequent purchaser.

British Library Cataloguing in Publication Data

Ikeda, Daisaku
Buddhism and the Cosmos
1. Cosmology, Buddhist
  I. Title   II. Kiguchi, Masayoshi
  III. Shimura, Eiichi
113'.0917'643   BQ4570.C6

ISBN 0-356-10773-6

Photoset in North Wales by
Derek Doyle & Associates, Mold, Clwyd.
Printed in Great Britain by
Redwood Burn Ltd, Trowbridge, Wiltshire
Bound at The Dorstel Press

Macdonald & Co (Publishers) Ltd
Greater London House
Hampstead Road
London NW1 7QX

A BPCC plc Company

CONTENTS

I	The Outer World of the Universe and the Inner World of Mind	9
II	What Is the Fundamental Law of Man and the Cosmos	25
III	Investigating the Mysterious Laws of the Universe	51
IV	Does E.T. Really Exist in the Universe?	82
V	Buddhism, the Universe and Human Existence	102
VI	Approaching Death's True Aspect from the Standpoint of Buddhism	142
VII	'The Threat of Existence' and the Mission of Buddhists	177
VIII	Life and Death: the Last Frontier	200
IX	The Life of a Star	217
X	The Potential for Life and the Concept of *Ku*	240
XI	The Nuclear Threat and the Buddhist Philosophy of Peace	266
XII	Does the Universe Undergo Birth and Death?	293
XIII	Expansion of Man's Perspective of the Universe	320
	Glossary	337
	Index	359

Foreword

In this volume, a comprehensive discussion unfolds, not on a global but rather on a cosmic scale, as an encounter between the universe – the eternal enigma of physics – and Buddhism, a philosophy of human existence. *Buddhism and the Cosmos*, as its title suggests, challenges the task of clarifying the relationship between the human being's inner world and the great cosmos, and also attempts to substantiate important Buddhist teachings in the light of the remarkable advances of modern science.

Science investigates the phenomenal world in which we live and act, dealing with the objectively perceptible aspects of existence. As the scientific approach has been applied to a wider range of disciplines, our knowledge of the outer world has progressed from superstition and supposition to substantiated theory and verifiable fact. Buddhism, on the other hand, elucidates the truths of the human being's subjective realm, thereby drawing out the practical wisdom of life. In its search for truth, science relies chiefly on analytical and inductive reasoning, while Buddhist reasoning entails a comprehensive method of deduction and intuition. The authors of *Buddhism and the Cosmos* believe that Buddhism and science have complementary roles to play in the achievement of human happiness. They express their conviction that the theoretical understanding offered by science lends credence to the Buddhist concept of the universe, and that the more science develops, the more fully the validity of the Buddhist view will be demonstrated. It also asserts that the Buddhist perspective can enable one to elevate his consciousness infinitely, awakening him to the urgent need for world peace and the sanctity of all other beings.

The three-way discussion presented here was originally

serialized in the Japanese monthly journal *Ushio* between May 1983 and May 1984. It was then published as a three-volume book entitled *Buppō to uchū o kataru* (*Buddhism and the Universe: A Dialogue*). The three participants are: Daisaku Ikeda, president of the Sōka Gakkai International (SGI); Masayoshi Kiguchi, member of the International Astronomical Union and lecturer at Kinki University; and Eiichi Shimura (moderator), editor of *Ushio* magazine. The SGI is one of the largest Buddhist cultural organizations in the world with member organizations in more than one hundred countries. Mr. Ikeda is not only an inspiring religious leader but very much a man of action, continually meeting and talking with ordinary people as well as political and intellectual leaders in his belief that understanding among nations begins with person-to-person communication. Dr. Kiguchi specializes in nuclear astrophysics, and Mr. Shimura is well versed in the literary realm. Their dialogue accordingly encompasses a variety of topics, including science, philosophy, literature and history, while shedding light on the mysterious drama of 'Buddhism and the cosmos'.

The editors have attempted to make the Buddhist terminology used in this dialogue readily accessible to the non-Buddhist reader, and to present as lucidly as possible both Buddhist concepts and the participants' views of the universe. Notes and a glossary are provided for some of the Buddhist or technical terms that appear in the text. In publishing this volume, we are indebted to the editorial staffs of the Ushio Publishing Co., Ltd., of Nichiren Shōshū of America, and of the Nichiren Shōshū International Centre.

I

The Outer World of the Universe and the Inner World of the Mind

Approaching the universe, endlessly expanding outwards, and consciousness, deeply expanding inwards, in mysterious space-time.

'The Starry Heavens Above and the Moral Law Within'

SHIMURA: Recently, the topic of the universe has excited tremendous interest. I believe it was the year before last when the American astronomer Carl Sagan and others produced a television series on this topic, which became quite popular.

KIGUCHI: This topic presents quite a challenge. However, it will be educational for me as a scientist, and I believe it will prove extremely significant to discuss even a bit the relationship between the human being's inner world and the great cosmos.

IKEDA: With respect to that grand stage in time and space that is the boundless cosmos, I am of course no astronomer but merely an interested layman. As a Buddhist, I find the topic extremely interesting because of its unfathomable vastness and depth. Nevertheless, the topic fascinates me, and if it proves in the least illuminating, I believe our discussion will be valuable.

SHIMURA: Even in the literary realm, the challenge of the universe – which is the original romance of science – has recently opened up a new field of scientific reading matter, such as an account of astronauts' experiences, in addition to science fiction.

KIGUCHI: And, moreover, the origin of life on Earth can in

every respect be considered in terms of its connection with the larger universe, so it is only natural that this topic receive attention.

IKEDA: The emergence of the human being has, in simple terms, been considered within the stream of evolutionary thought propounded by Darwin. And the origin of life itself may have followed a similar evolutionary process. Volcanic eruptions spouted from beneath the earth and precipitated as rain through the action of water vapour. The rain formed rivers and, eventually, seas. Lava, compounds in the atmosphere, and the like, dissolving in the boiling seas, may have helped to form the first organic molecules, and thus served as a contributing factor in the origin of life. There is also a theory that the raw materials, so to speak, for the emergence of life may have been generated by lightning.

KIGUCHI: Actually, that theory based on evolutionary thought is now considered rather antiquated. The present trend is to grasp the birth of humanity within the same scheme as the evolution of the universe and the formation of the Earth. For example, one theory maintains that repeated collisions of comets with the solid body of the Earth once it had cooled somehow contributed to the emergence of life. Another theory has it that oxygen and nitrogen trapped in the earth spontaneously broke free and formed the atmosphere, essential to life's appearance.

IKEDA: What you have just mentioned about the question of life's origin calls to mind a discussion I had the year before last (1981) with Dr. Anatoliy A. Logunov, rector of Moscow State University. He, too, seemed to think – to state it simply – that life emerged on the primordial Earth in the course of the process of chemical evolution occurring in interstellar space, and began to develop from there.

KIGUCHI: Alcohol and other organic compounds actually are being discovered in interstellar space. There is also a theory which regards comets and meteors as the source of the raw materials of chemical evolution.

SHIMURA: Some scholars claim that primitive life originated from the heads of comets, though this is a minority opinion. All in all, I think we can say that a definitive explanation has yet to be established.

IKEDA: The prevailing theory concerning the origin of the universe itself is that it began with the Big Bang. But was this Big Bang the beginning of everything, or did the universe exist before that in a phase of contraction? We find a great divergence of opinions on this point.

KIGUCHI: That question is still under investigation, and not even an expert can say for certain.

SHIMURA: Mr. Ikeda, I would like to ask what you think about these numerous mysteries surrounding the universe, especially from the standpoint of Buddhism.

IKEDA: The topics involving the universe and life itself are incredibly vast in scope. Even speaking from the standpoint of our present scientific development, as is evidenced, for example, in the success of the *Apollo* lunar voyages, I feel we have arrived at only the most fragmentary understanding. But the field of astronomy seems to be advancing at an astounding pace. Who are some of the individuals in the forefront of the field?

KIGUCHI: It is true that astronomy is an ancient discipline, but it is actually a science of the future. In our country, after the Second World War, Professors Hideki Yukawa and Kōji Fushimi declared its importance, and after that, many outstanding astronomers appeared, including Chūshirō Hayashi (professor at Kyoto University), Sachio Hayakawa (professor at Nagoya University), and Minoru Oda (professor at Tokyo University). Their students are now active in the field.

Among the world's great astronomers outside Japan, we have, in the United States, Subrahmanyan Chandrasekhar (professor at the University of Chicago) and John A. Wheeler (professor at Princeton University); in Denmark, we have Bengt G.D. Strömgren (honorary professor at the University of Copenhagen); and at the observatories, Professor Allan R. Sandage at Mount Palomar, to name just a few. Also, Dr. Carl Sagan is currently active in the study of the solar system.

SHIMURA: What sort of reference works are used in astronomy?

KIGUCHI: Well, we read the latest research, much of which is written in English, as well as works that are considered classics in the field, such as Chandrasekhar's *Introduction to the Study of Stellar Structure*, or the British astronomer Sir Arthur S. Eddington's *The Mathematical Theory of Relativity*. Another one that comes immediately to mind is Sandage's photo collection, *The Hubble Atlas of Galaxies*, which is very useful in galactic studies.

IKEDA: I feel we should expect great things from the knowledge gained through astronomy.

KIGUCHI: As a scholar in that field, I most earnestly hope we can reply to such expectations. However, even if we fully extend the wings of speculation based on the knowledge science has accumulated so far, and on the conjecture of scientists, our view of the vast cosmos still does not extend beyond the merest and most fragmentary theoretical postulation.

IKEDA: I understand. I don't recall exactly when, but during a discussion I had with the Japanese writer Yasushi Inoue, he mentioned to me something Immanuel Kant once said. Your remarks reminded me of it. Kant wrote, 'Two things fill the mind with ever new and increasing admiration and awe, the oftener and more steadily we reflect on them: the starry heavens above and the moral law within.' It is only a brief statement, but it never fails to move me.

KIGUCHI: I've heard that, during a lecture at his alma mater, Kant said something to this effect: 'I do not expect you to learn philosophy from me. Rather, I hope that you will study philosophy through self-contemplation.' It seems he wanted his students to penetrate the truth through increasingly deep introspection.

IKEDA: Kant's scientific awareness apparently derives from Newton. We may say that Einstein was the one who finally broke free from the bounds of Newtonian physics. However, setting aside for the moment Einstein and his contribution to modern physics and astronomy, one can sense a profound significance in these words of Kant. The genius of this expression lies in its juxtaposition of the eternal and boundless cosmos, which even human reason can perhaps never fully grasp, and the inner realm of the mind. One cannot help but be impressed by this superb contrasting of the heart or mind and the universe. In short, while the universe is a vast and mysterious expanse in time and space, the human mind too unfolds into a world – a universe – of endless subtle changes. On the one hand, we have an infinite outward expansion; on the other, an expanse of unfathomable inner depths; and at the same time, these two are linked together. The interrelation of the outer cosmos and the inner realm of mind is clarified in Buddhism with the principle of *ichinen sanzen*.

The Interlocking Nature of the 'One Mind' and the Universe

SHIMURA: We find many religions in the world. However, I believe – to paraphrase the late Arnold Toynbee – that among the higher religions, it is the teachings of Mahayana Buddhism, specifically those based on the *Lotus Sutra*, which hold a crucial answer for the future. Mr. Ikeda, I believe you discussed this point with him on several occasions.

IKEDA: Yes, I did. Shakyamuni Buddha, who made his advent nearly three thousand years ago in India, expounded the twenty-eight-chapter *Lotus Sutra* as the pinnacle of his so-called eighty thousand teachings. The 'three assemblies in two places' appearing in that sutra may be viewed as clarifying the relationship between the 'one mind' (*ichinen*) and the universe.
Here this discussion may become a bit complex, but in any

event, the 'three assemblies in two places' is a division of the *Lotus Sutra* according to the sequence and location of the events described in it. It refers to Shakyamuni's initial preaching to the assembly on Eagle Peak; the Ceremony in the Air, during which he lifts the entire assembly into space and continues his preaching; and his final return of the assembly to Eagle Peak where he concludes his preaching.

In particular, the Ceremony in the Air depicts the essence of a magnificent view of life and the universe expressing the boundlessness of the inner realm of the life-moment, and linking that life-moment with the vast and limitless cosmos. The Great Teacher T'ien-t'ai of China systematized this view as the principle of *ichinen sanzen*, or three thousand realms in a single life-moment.

The word *ichinen*, literally 'one mind', is used in Buddhism to mean the ultimate reality of life at each moment, a usage quite different from its ordinary sense of 'single-mindedness' or 'determination'. Buddhism grasps the true aspect of life at each moment minutely and logically with the doctrine of the Ten Worlds. Life at each moment contains within itself Ten Worlds, a hundred worlds, a thousand factors and three realms, consisting of oneself, the environment, and other living beings. Through the interrelation of all these, the life-moment expands into three thousand realms – a very clear and logical analysis.

There is an unforgettable passage from Nichiren Daishōnin's writings which concerns the view of life I have been discussing. He states, 'When one examines the nature of life, he will find no beginning which necessitates birth and no end which requires death; rather he will discover the true entity of life which is free from birth and death. This entity of life cannot be consumed by the flames at the kalpa's end,[1] nor can it be washed away by floods. It cannot be cut by swords, nor shot by arrows. Although it can fit inside a mustard seed, the seed does not expand, nor does this life contract. Although it fills the vastness of space, space is not too wide, nor is life too small.'

Buddhism as Deductive, Science as Inductive Reasoning

KIGUCHI: What a fascinating passage! As a scientist, I can definitely sense its scope. This must be because the logic of Buddhism is *deductive*, while that of science is *inductive*.

IKEDA: You are quite right. Science and Buddhism in no way contradict one another. I believe that as science progresses, the understanding of Buddhism will advance, and that Buddhism in turn can provide scientists with an inexhaustible source of fuel for thought. A feeling of mutual respect and sincerity is needed, wouldn't you say?

SHIMURA: I quite agree. What you say calls to mind Einstein's words, 'Science without religion is lame, religion without science is blind.'

IKEDA: My teacher, the late Jōsei Toda (1900-1958), once heard Einstein lecture when he visited Japan in November 1922. He often described that occasion to me. It would seem that the words of someone who has thoroughly investigated a particular truth have a certain depth. They penetrate the heart, and one cannot forget them as long as he lives.

I am sure I will have the chance to touch on Einstein's theories in more detail later on, but for now, let us return to Kant's statement. While he contrasts the depths of the mysteries hidden in the universe with 'the inner realm of mind', he does not go beyond the contradistinction of the two. Herein his understanding differs essentially from that of the Buddha, who grasped the universe from the standpoint of his enlightenment.

However, I would like to point out that, transcending philosophical speculation, Kant says that he was moved to 'admiration and awe'. I think such humility on the part of Kant, a man of no ordinary intellect, is indeed splendid. In this connection, we ourselves must not stop at merely contemplating the mysteries of the universe. Herein, I believe, lies the necessity of the actual feeling or experience derived from the faith, study, contemplation and action of Buddhism.

Simply to ponder the universe while avoiding the mundane

realities of human existence and society would be to degrade the whole subject to a stage for our curiosity. Our desire for abstract knowledge might be satisfied, but that would be all. For this very reason, I believe that one should strive to grasp the relationship between himself and the vast expanse of the cosmos while embracing Buddhism, expanding his moment-to-moment state of life in the context of daily affairs and overcoming whatever troubles might arise. This, I believe, is the proper approach for us to take.

KIGUCHI: I agree with you wholeheartedly. One should not divorce the universe from human existence. Incidentally, among many peoples of both East and West, we find the belief that the phenomena in the heavens are related to, or are a relation of, human affairs.

SHIMURA: The idea that the position of the constellations exerts an influence on human destiny, or that man's fate is reflected in the stars, has been well established in certain eras.

IKEDA: One finds many references to such supposed relationships, for example, in the Chinese classic, *Romance of the Three Kingdoms*. One well-known example is the scene where Chung-ta (Ssuma Yi), observing the heavens, infers the death of his enemy Chu-ko K'ung-ming. Actually, K'ung-ming outmanoeuvred him. Knowing that Chung-ta would read his death in the stars, he made his last testament and then commanded his men to behave, even after his death, as though he were still alive, thus causing Chung-ta to fall into a trap. This incident is quite famous.

In the West, the origins of astrology can be traced back at least to ancient Babylon, but it is said that precise, systematic explanations of the art date from the Renaissance and after. Even now, though they no longer regard it as a body of arcane secrets, many people still believe to some extent in astrology.

KIGUCHI: Science still cannot shed much light on fundamental problems of life and death. For example, we do not know when we will die, and the question of where life comes from remains shrouded in mystery. Whether or not there actually exist somewhere in the cosmos other living creatures like human beings is another issue which has yet to be clarified. Even from the standpoint of astronomy, the pursuit of answers to such questions will surely involve vast amounts of time. In this connection, I believe that Buddhist deductive thought offers us an important key.

The View of the Universe

IKEDA: Buddhism at the outset discerned the nature of birth and death, and sought to reveal the way to overcome the sufferings that these entail. Moreover, I believe it clearly teaches how one can lead a peaceful life while perceiving the true aspects of birth, old age, sickness and death just as they are, and also elucidates the nature of one's subjective existence that 'melts' back into the cosmos during the interval between death and rebirth in the eternal cycle of transmigration. In short, I believe we can say that Buddhism does not analyze the universe itself in this or that way but rather clarifies the ultimate truth or Law permeating it — a truth which is unadorned, without beginning, and naturally inherent. It describes this Law, for example, as 'uncreated' (*musa*) and as 'that which cannot be defined as either good or evil' (*muki*).

Among the great thinkers of ancient Greece and the Orient, and among the scholars of ancient India, we find many people of remarkable knowledge who devoted themselves to astronomical observation. Shakyamuni, however, dealt with a dimension different from that of astronomy. First he raised the question, 'How should human beings live?' and came squarely to grips with the problem of human suffering, represented by the four universal sufferings of birth, old age, sickness and death. What he awoke to as the result of his endeavours was the ultimate truth of the 'inner realm of mind'.

SHIMURA: But even in the teaching of Buddhism we find a definite cosmology or world view, as expressed, for example, in such concepts as the 'ten directions' or the 'four continents surrounding Mount Sumeru'.

IKEDA: That is quite true. The sutras contain numerous instances of reference to the cosmology or world view of ancient India, derived from astronomical conceptualizations of the time, which were used to explain the Buddha's enlightenment. However, I do not think this cosmology is central to the Buddhist sutras. Their reference to the heavens never deals, for example, with problems of the dimension illustrated by our previous example from *Romance of the Three Kingdoms*. Nevertheless, we cannot help but be surprised at the wealth of imagery and the depth and breadth of insight implicit in the view of the world and the universe to be found in the Buddhist sutras. It appears that on some points the thinking of the ancient Indians was fundamentally consistent with the findings of modern astronomers and theoretical physicists.

KIGUCHI: Then, to expand on your earlier remarks, the prevalent view of the universe was used as a sort of introduction to Shakyamuni's teachings about the solution of the fundamental problems of birth, old age, sickness and death, is that correct?

IKEDA: Yes, indeed.

KIGUCHI: Even so, it is astounding that the sutras should be fundamentally consistent with modern astronomy or theoretical physics.

IKEDA: It is. However, I wish to make clear that what Buddhism takes as the object of its penetrating study is 'the inner realm of mind'. It considers the world and the universe in terms of their connection with the human mind. That is, even when Buddhism discusses the universe, it grasps it chiefly as a means, when viewed *subjectively*, to think about one's own mind, and when viewed *objectively*, to think about one's daily affairs or way of living. This approach is basic to the Buddhist view of the universe – a point I think we would do well to remember.

SHIMURA: I understand. To turn and gaze again at the human

being with eyes that have beheld the universe is also, albeit in a non-Buddhist sense, the stance Dr. Sagan has adopted.

KIGUCHI: I noticed that, too. For example, in the conclusion to *Cosmos*, he states, 'For we are the local embodiment of a Cosmos grown to self-awareness. We have begun to contemplate our origins: starstuff pondering the stars; organized assemblages of ten billion billion billion atoms considering the evolution of atoms; tracing the long journey by which, here at least, consciousness arose ... Our obligation to survive is owed not just to ourselves but also to that Cosmos, ancient and vast, from which we spring.'

Hideo Kobayashi and Death

SHIMURA: We have now heard that the basic position of Buddhism lies in its attempt to solve the four universal sufferings of birth, old age, sickness and death. But in Japan in recent years, apart from religious figures, has any outstanding individual or book directly addressed the question of life and death?

IKEDA: Well, there is the literary critic Hideo Kobayashi (1902-1983). In 1971, I had a leisurely discussion with him over lunch at the Ren'yōan at our head temple, when the cherry trees were in bloom. Ton Satomi and Mitsuo Nakamura were also with us.

I feel that in the transition of his thought and views, Kobayashi was closely approaching T'ien-t'ai's principle of *ichinen sanzen*. His progress in stages through such subjects as Confucianism and Shinto (that is, the Shinto of the *Kojiki* [*Records of Ancient Matters*] or Motoori Norinaga) seems to offer us clues to the evolution of his thinking. It is to be greatly regretted that we have lost such a truly wise person.

Several days after we met, I sent him some cherry blossoms as a token of my appreciation. I was told that at that time, among other things, he said, 'I have been preparing for death since I turned sixty.' When I heard this, I suddenly realized that he was even then striving to confront death directly.

SHIMURA: Recently I happened to hear that when Mr. Kobayashi was near death, he told someone who had come to visit him that he felt rather as though death had counter-attacked him.

IKEDA: Naturally, he could not challenge and overcome death by reason alone. Dr. Toynbee had a similar experience, I believe. He had pondered death with the power of his intellect, but when death actually confronted him, he was forced to realize that the problem of death is precisely that one cannot overcome it in the intellectual realm.

The Role of Gamow's Theory of the Universe

KIGUCHI: Kant coined the word 'nebula', didn't he?

IKEDA: So I've heard. We might say that even Newton first came to be appreciated in a broad academic context through the medium of Kant's philosophy. By means of Newton's dynamics (which involves the principle of universal gravitation), Kant attempted to consider a vast range of subjects, from the formation of the universe to a fundamental understanding of the human being.

SHIMURA: That is to say, the establishment of Kant's philosophy simultaneously brought Newton's dynamics to public attention?

IKEDA: Not only that, but I believe Kant's philosophy can be understood as a step toward the view of the world or universe expounded as the ultimate realization of Buddhist enlightenment, in which the 'self' is coextensive with the universe (*ga soku uchū*) and the universe is inherent in the 'self' (*uchū soku ga*).

KIGUCHI: This may be a rather difficult topic, but in the latest development of theoretical physics we find an interpretation of the theory called the 'anthropic principle', which has intrigued many physicists. In essence, it states that because our own life exists, the universe also exists.

SHIMURA: That's very interesting. Simply stated, exactly what is this interpretation and what form will it take from now on in theoretical physics and related fields?

KIGUCHI: Briefly, it states that the universe is a collection of various potentials or possibilities, of which human beings are only one. According to this way of thinking, even our present universe itself may be only one of many other possible ones. To give an example, in the realm of elementary particles studied by Dr. Hideki Yukawa and others, it has been found that there is a close link between the existence of elementary particles and the fact that human beings are observing their phenomena. I believe that, even as applied to the vast scale of the universe, this interpretation will be perfected in the future as a theory based on proper data.

IKEDA: It would seem that Gamow's theory must have played some role in this interpretation of the 'anthropic principle'.

KIGUCHI: You are quite right.

The Neglect of the 'Inner Consciousness'

SHIMURA: Gamow visited Japan once.

IKEDA: Ah, yes, in the autumn of 1959, I believe. Unfortunately, his visit coincided with the typhoon in Ise Bay, and I was so busy working to aid and encourage the relief effort that I didn't even have time to carefully read the newspaper article reporting his lecture. However, I fondly recall that much earlier, as a young man, I often used Gamow's textbook (translated into Japanese by Kōji Fushimi) when I studied each morning with Mr. Jōsei Toda, himself a mathematician.

Gamow's theory of an expanding universe must have stemmed from thinking about space as a limitless immensity.

Einstein himself, in pondering the vast reaches of the cosmos, must have been led to contemplate the infinite expanse of his own inner consciousness that could ponder this subject. After repeated thought, he must have used his words 'cosmic religious sense' to express the awesomeness of the boundless world of the mind.

I think that Gamow's theory of an expanding universe may have been inspired, at least in part, by the theories born of Einstein's most profound speculation on the outer cosmos and the inner realm of mind. That is to say, unless one possessed a limitless expanse within himself, he could not conceive the idea of an endlessly expanding universe.

KIGUCHI: In the academic world this became a difficult and specialized theory, but your words just now touch on the core of the thinking central to that theory's structure.

The idea that a limitless 'inner realm of mind' exists along with the boundless spatial expanse of the 'outer cosmos' has, to some extent, become generally accepted in the realm of modern theoretical physics and particle engineering.

Gamow's theory of an expanding universe was postulated in the 1940s, and a little more than ten years ago significant evidence was obtained in support of this idea. This data was gathered by two electronics engineers working at Bell Laboratories in the United States, who won the Nobel Prize for their research.[2]

In any event, we can say that this theory has been developed by observation, experiments and data. Moreover, with the development of the radio-telescope we can now see up to a distance of three billion light-years. And since there are more than a billion galaxies, the number of stars we can calculate from that amounts to more than a trillion. The universe is vast indeed.

IKEDA: However, compared to our investigation of the 'outer cosmos', our understanding of the 'inner realm of mind' lags far behind. The challenge of the external universe has brought about amazing progress in scientific technology, which has in turn led to the flowering of our civilisation and the prosperity of our modern life. From computers to instant foods, and from robots to clones, scientific achievements are greatly altering our way of living. Yet because our understanding of the 'inner realm of mind' has been neglected, human beings, who should by rights play the leading roles, are forced to play a subservient role in science. Recently this feeling has been growing stronger and stronger.

SHIMURA: Apparently everyone shares this feeling. While at first glance our age seems to be flourishing, a profound uneasiness lurks in people's hearts.

The Desire for Cosmic Religious Sense

IKEDA: This uneasiness might be called a characteristic of twentieth-century *fin-de-siècle* awareness.

As is well-known, in the late nineteenth century – especially in Europe, regarded as the centre of civilisation – Christianity, the keeper of man's inner consciousness, came under attack from Nietzsche and others who proclaimed that 'God is dead'. No doubt this declaration served as a metaphor for the bankruptcy of Christian theology. On the other hand, no sooner had man been released from God's spell, so to speak, than the academic world, which had been bound by the Christian world view, science in particular, made truly astonishing progress. However, there was no one at God's deathbed to inherit the responsibility for guiding the inner realm of the human mind.

SHIMURA: Einstein, pondering this problem, may have conceived his phrase, 'cosmic religious sense', to convey a desire that arose from thinking about the image of that inheritor.

IKEDA: That's quite likely. We can say that on the whole he was not only an outstanding scientist but an outstanding thinker.

Einstein, who had had no opportunity to study Buddhism, produced only this abstract and rather vague phrase from the culmination of his speculation as a scientist, but his 'cosmic religious sense' calls to my mind what we mean by the term the '*ichinen* of prayer'.

To change the subject – Dr. Kiguchi, the world is said to be round, isn't it? Or slightly obloid, as some people say? Are the other stars and their planets – more than one hundred billion of them – also round?

KIGUCHI: We cannot say that they are. Some stars appear disc-shaped, or even square. Of course, these shapes are probably due to the distortion caused by light refraction.

However, most stars can probably be said to be round. A round shape has the greatest density – that is, it is the shape by which it is easiest for an object to support its own weight.

IKEDA: I see. In Buddhism, too, we find the phrase *en'yū enman* (completely round and perfect), which describes the ideal of a perfected human character.

KIGUCHI: There is also the Gauss theorem. While physicists find this a rather elusive principle, in simple terms it states that, in assuming a round shape, weight becomes distributed most effectively, and this shape is least vulnerable to outside forces.

IKEDA: Such principles or theorems always conform to reason. Indeed, they cannot be irrational. Buddhism also has its own reason or logic. The sutras – the Buddha's teachings – contain explanations of the Law of life that conforms to the great Law of the universe.

NOTES: CHAPTER I

1. The kalpa's end: the end of the world when, according to the Buddhist sutras, the three major calamities will occur (of fire, water and wind).
2. Arno Allan Penzias, astrophysicist and Vice President of Research, Bell Laboratories, and Robert Woodrow Wilson, radio astronomer and member of the Technical Staff of Bell Telephone Labs. They won the Nobel Prize for Physics in 1978.

II

What Is the Fundamental Law of Man and the Cosmos?

Considering the interrelationship of the cosmos and human life by exploring the depths of the object of worship in prayer, what is the most precious human act?

On the Significance of 'Prayer'

SHIMURA: During our first discussion, we approached our subject in overall terms. Now, how about discussing specifically the relationship between the Buddhist Law and the universe? Then I think it might be interesting to take up topics such as *E.T.* − a movie now much talked about the world over − and peace in the universe. Also, I would like to ask the two of you to discuss the problem of birth and death, and the cycle of formation, continuance, decline and disintegration (*jō jū e kū*) as taught in Buddhism.

KIGUCHI: In our last session, Mr. Ikeda, you mentioned the *ichinen* or 'one mind' of prayer in connection with Einstein's phrase, 'cosmic religious sense'. I think the topic of prayer is of primary importance in religion.

SHIMURA: There is an interesting story in connection with 'prayer'. A number of Japanese man-made satellites have been launched from Uchinoura in Kagoshima Prefecture. On one occasion, although all preparations had been completed, launching was postponed due to bad weather. When it became evident that any further postponement would mean abandoning the launching project, scientists of the then Tokyo University Institute of Space and Aeronautical Science, as well as project-member scientists representing other parts of the country, went to offer their earnest prayers at a nearby Shinto shrine.

IKEDA: The attitude of prayer in Buddhism is expressed as *namu* or *nam* which derives from Sanskrit. *Namu* is translated into Japanese as *kimyō* and means 'to devote one's life'. This act of devotion might be described as a ceremony in which one fuses one's own being with the fundamental Law of the universe itself. In other words, Nichiren Daishōnin, the Buddha of the Latter Day of the Law, embodied the fundamental Law of the universe in the form of a scroll or mandala. In one's prayer to that mandala, the act through which devotion takes place, the external cosmos and the inner realm of mind are perfectly fused, enabling one to establish a correct rhythm in the course of life and daily existence.

KIGUCHI: I understand that very well. Two hydrogen atoms combine with one oxygen atom to form water. But a certain substance which never changes is needed to act as an intermediary. Unless something like platinum black is employed as a catalyst, no water is in fact produced. Though of course the dimensions differ totally, we could say that, in simple terms, the object of worship serves as a sort of catalyst to bring about the fusion of the individual's subjective mind and the objective reality of the universe in the attainment of enlightenment.

IKEDA: From this perspective, the scientists' visit to the shrine was apparently intended as a prayer to the god for help in a time of trouble – an act on a level rather different from that of Einstein's 'cosmic religious sense'.

SHIMURA: By mandala, I take it you mean what we generally call *honzon*, or object of worship?

IKEDA: That is correct. *Honzon* means that which is respected as fundamental. It is also one translation of the Sanskrit *mandala*. Some others are 'cluster of blessings' and 'perfectly endowed'. Generally speaking, the act of prayer serves to fully activate the inmost depths of the human spirit. Animals are not endowed with this capacity. Prayer is indeed the noblest of all human acts. For that reason alone, Buddhism places great emphasis upon the object to which prayer is directed. An infinite variety of objects of worship are to be found, differing in dimension and depth. Muslims offer prayers to Allah – the omniscient and omnipotent creator of the universe – saying that Allah is great. In Judaism, Yaweh or Jehovah is worshipped as the transcendental God.

SHIMURA: Feuerbach asserted that 'God' in Christianity was merely a conceptual construct and thereby denied His existence. Instead, he contended that God is an outward projection of man's innate nature.

IKEDA: I would like to save this topic for another occasion and return to the subject of prayer. We find that prayer takes many forms. In his teaching, Nichiren Daishōnin declared that one should take as the object of worship that which is supreme, unrivalled and fully supported by documentary, theoretical and actual proofs.

SHIMURA: Take for example the Hachiman Shrine, sacred to the god Hachiman. How does Buddhism view Hachiman?

IKEDA: All existences in the universe are embodied in the Gohonzon, the object of worship in Nichiren Shōshū Buddhism. In ancient Japan, Hachiman was worshipped as the god of weaving or as the protective deity of the harvest. According to one decidedly Buddhist interpretation, *hachi* (eight) in Hachiman is construed to mean the eight scrolls of the *Lotus Sutra*. Moreover, the character *man* has the cloth radical, signifying clothing, and the remainder of the character is written with the symbols for 'rice' and 'field', indicating rice and other grains. The character *man* has thus been taken to mean 'conferring the benefits of food and clothing'. Hachiman is depicted on the Gohonzon as one of the many benevolent deities of the universe. This indicates that Hachiman is a manifestation of the beneficent power inherent in one's own life.

On the Gohonzon are depicted the god of the Sun (Dainitten), the god of the Moon (Daigatten), the god of the stars (Daimyōjōten) and other deities, who represent the life existing in the outer cosmos, the movement of heavenly bodies, and the workings that bring about harmony.

If one's prayer is vague, there will be no profound response between the outer cosmos and the inner universe that is one's own life-moment.

KIGUCHI: In nature worship or shamanistic practices, prayer and its objects seem to be shallow and superficial.

IKEDA: Prayer in Nichiren Daishōnin's Buddhism means to devote oneself to the Gohonzon, the ultimate entity of that which is to be respected as fundamental. The *Ongi kuden* (*Record of the Orally Transmitted Teachings*) states, '*Namu* derives from Sanskrit and means to devote one's life.' The Sanskrit *namas* is interpreted in Buddhism as 'devotion of one's life'. Other translations are to honour, to revere, to return to, to respect, to seek salvation, to attain the Way, and so forth. Be that as it may, our *ichinen* in the moment of prayer permeates all our human activities constituting the three categories of action: our mental, verbal and physical actions. That is why the object toward which prayer is directed becomes crucial. The mandala or Gohonzon inscribed by Nichiren Daishōnin is the 'cluster of blessings'. Therefore it is only natural that, as the *Kitō shō* (*On Prayer*) states, '... it could never happen that the prayer of a votary of the *Lotus Sutra* would go unanswered.'

In any event, as Mr. Shimura's story indicates, science has its limits. Science – whether in astronomy or in any other field – has from the outset challenged the external world, from the origins of life to the boundless, mysterious realms of time and space. I feel that in studying the universe from now on, it will be necessary to exert the same intensity of effort as that needed to pursue and fathom life's inner realm.

KIGUCHI: Scientific study of the universe has indeed

advanced, yet we must realize that its progress has served to establish the fact that there are actually two realms: one that science can clarify and one that it cannot.

IKEDA: It seems to me that the realm science cannot explain is that pertaining to life itself.

Our present society tends to place even the highest teaching of Buddhism – which does clarify this mysterious realm – in the same category as various other religions, dismissing it as irrational and not worth bothering about. Obviously this is a gross mistake. Arnold Toynbee and other scholars of outstanding intellect, men who have contributed greatly to the betterment of humanity, have come to direct their studies toward the realm of the higher religions.

I believe that we have entered an age when the higher religions, including Mahayana Buddhism, must be seriously investigated. I am convinced that as long as people remain unaware of this need, a full awakening to our human potential and the perfection of culture cannot be achieved.

Peace and Happiness within Human Life

SHIMURA: These days, even in the mass media, we find frequent discussion of the term 'analogy' – here indicating a recognition of universal existence transcending the domain of science – as well as such topics as 'the age of intellectual reform', an expression pointing to the demand of our times for a deeper understanding of the human consciousness.

KIGUCHI: We have indeed entered an age when 'intellectual reform' has become crucial. Moreover, I believe that reference to a universal truth transcending the realm of science, as indicated by the expression 'analogy', is definitely valid. So far, however, intellectuals have hit upon no concrete method of approach, and seem to be wandering in a vast labyrinth.

IKEDA: I agree with you. The Law inherent in the universe since time without beginning, as well as principles underlying electro-magnetic waves, gravitational waves, and other natural phenomena, is all inherent within the individual human life. Because of this, we may say that human beings have been able to clarify numerous phenomena by means of science, and that this has in turn provided the basis for numerous practical applications and inventions. In short, no principle or law exists independent of or separate from one's own life; all exist within the context of their interrelationship. This, I believe, is a correct understanding. Nichiren Daishōnin's *Isshō jōbutsu shō* (*On Attaining Buddhahood*) clearly states that the Law is not outside one's own life.

Buddhism expounds this interrelationship between the individuals and the cosmos as the principle of *eshō funi*, or the oneness of life and its environment. Here *eshō* is a contraction of *shōhō*, the subjective individual, and *ehō*, his objective surroundings. These two exist in a relationship of 'oneness' or non-duality (*funi*).

In other words, the laws and principles inherent in the universe since time without beginning also exist in the depths of one's own life, and for that very reason, the human race has been able to uncover these principles, making various discoveries and inventions. It is only natural, therefore, that the fruit of these achievements be used for the peace and happiness of humanity.

SHIMURA: It is appropriate to view all universal laws as inherent in the human *ichinen* or life-moment, isn't it?

IKEDA: I believe so. The Buddhist teachings are clear and specific on this point. Nichiren Daishōnin's *Sanze shobutsu sōkanmon shō* (*On the Teachings Affirmed by All Buddhas throughout Time*) states, 'One's body imitates in detail the heavens and the earth.' In expounding the principles that the individual life permeates the universe and that the universe is inherent in the individual life, this passage explains their interrelationship with easily understandable analogies, drawing correspondence between the cosmos and the human body.

For example, it likens the roundness of the head to the heavens, the feet to the Earth, and the body cavity to space. It compares the belly's warmth to spring and summer, and the back's hardness to autumn and winter, thus correlating the body with the four seasons. It also likens the 'four regions' – head, arms, trunk and legs – to spring, summer, autumn and winter. The three major joints of the hands and feet are together compared to the twelve months, and the 360 minor joints, to the days of the year. Further, it likens breath passing in and out through the nose to wind blowing through the hills and valleys, and breath passing in and out through the mouth, to wind blowing through the vast sky.

SHIMURA: Then catching a cold would be analogous to a typhoon. Tornadoes and hurricanes might be thought of as nature's illnesses ... Joking aside, I think this passage sheds light on the relationship of the human being with the natural environment.

IKEDA: I believe we can say that. It further likens one's two eyes to the Sun and Moon, and their opening and closing, to the phenomena of day and night. The hair of one's head is likened to the stars and constellations; one's eyebrows, to the Big Dipper; the flow of one's blood, to rivers and streams; one's bones, to jewels and stones; one's skin and flesh, to earth and soil; and one's body hair, to a forest of luxuriant trees.

KIGUCHI: Soil becomes depleted, and plants and trees wither, just as one's hair thins with the passing years.

IKEDA: The *Sōkanmon shō* also likens the 'five major organs' – the lungs, heart, spleen, liver and kidneys – to the 'five stars in heaven', which we may take to mean Mercury, Venus, Mars, Jupiter and Saturn.

KIGUCHI: These are truly fertile images, springing from intuitive wisdom.

IKEDA: The divorce of science from the all-important concern of human life, that which motivates people to seek out truth – whether one calls it intellect, curiosity, or imagination –

has become a truly pitiful and distorted thing. For this reason, even though science may have departed from the interests of the human being, it will inevitably have to return to benefit him. Out of this process, I believe, a correct view of both science and life will emerge.

KIGUCHI: We may say that space studies have grown up amid the life sciences. For example, there are planetary ecology, which views humanity and the Earth itself as part of the cosmic ecosphere; bionics, which applies the superb functions of living organisms to the context of engineering; and bioharmonics, which seeks to harmonize life-activities. Therefore, scientists who study space do not have nearly the resistance that lay people might imagine them having toward worshipping, with palms pressed together, a transcendent reality beyond human knowledge. In fact, during their 'danger years',[1] more than a few scientists go to offer prayers at Shinto shrines.

IKEDA: The prayers of those scientists at Uchinoura may have been an expression of their religious feeling, but viewing matters in a harsh light, I feel they probably offered those prayers simply because it is a customary thing to do.

SHIMURA: Or because they had no other alternative. The other day, the space shuttle *Challenger* was launched only 0.008 second (eight-thousandths of a second) later than planned. This was amazing precision. By the way, in the United States, the space exploration program monopolizes the top-level researchers, and it is said that as a result, the steel, home appliance and automobile industries are now stagnant.

KIGUCHI: On television one sees the footage of manned space laboratories – surprisingly not all that large – being loaded onto the space shuttle. No less than a thousand of the most advanced high-technology firms are participating in the development of the necessary equipment.

SHIMURA: In watching a televised transmission of the space shuttle launching, I was deeply impressed by the astronauts and the closely involved scientists who often used the word 'pray' in their responses to the interviewer, saying, for example, 'All that remains is to pray for success,' or, 'We pray for a safe return.'

IKEDA: The distance from the Earth's surface to the detectable limits of the atmosphere where outer space begins is about thirty thousand kilometres. In terms of distance on land this is about 4.5 times the length of the Nile. Until quite recently, outer space was regarded as an environment that could not support life.

SHIMURA: Yet they say it will be developed into an environment that *can* support human life. Modern science has successfully taken the first step toward that goal.

IKEDA: Yes, indeed. Viewed in terms of the history of evolution since the time of life's emergence, the advent of space flight could be regarded as an event that may or may not take place once in several billion years.

SHIMURA: Mr. Ikeda, you once referred to the twenty-first century as the 'century of life'. Did you mean that the twenty-first century will mark a great turning point in the course of human development?

IKEDA: First, of course, I meant that the new century must resolve certain fundamental problems involving life and death for humanity as a whole. In terms of our relationship to the cosmos, we have already entered an unprecedented space age, and I'm sure the twenty-first century will witness its continued advance.

KIGUCHI: I believe that our responsibility as astronomers will become crucial. What new understanding our discoveries in the field can bring to humanity and what role they will enable human beings to play will no doubt be a major key to the new age.

SHIMURA: Earlier, Mr. Ikeda, you referred to a passage from the *Sōkanmon shō* which reads, 'When one examines the nature of life, he will find no beginning which necessitates birth and no end which requires death.' Since this passage is rather difficult, would you kindly explain it in simpler terms?

IKEDA: Simply stated, our life – that is, the true nature of life at each moment – has neither beginning nor end, and shall not vanish from the world, or the universe, at death. Life itself possesses an inherent reality or dimension that is eternal, transcending birth and death. As the writing you mentioned states, it cannot be destroyed even by flames big enough to consume the world, nor can it be washed away by the most disastrous flood. It cannot be cut by a sword nor pierced by an arrow. Although it can fit inside the tiniest mustard seed, the seed will not expand, and although it fills the universe, the universe is not too vast. In short, the life-moment is an unchanging reality transcending all relative distinctions of birth and death, emerging and vanishing, large and small, broad and narrow. I read this passage as a testimony to the true nature of life.

KIGUCHI: That passage touches on the mysterious nature of life. Physics, too, recognizes a dimension that is 'absolute', in contradistinction to time and space, which belongs to the realm of relativity.

IKEDA: We may say that life is a reality having neither beginning nor end, and permeating both time and space. Buddhism defines this reality as the true 'self'.

SHIMURA: Couldn't we say that the clarification of this 'self', the core of life, so to speak, constitutes the essence of Buddhism?

IKEDA: That is correct. The *Muryōgi-kyō* (*Sutra of Infinite Meaning*), an introductory teaching to the *Lotus Sutra*, clarifies the nature of the 'self', saying, 'This entity is neither existence nor non-existence, neither cause nor circumstance, and neither self nor other. It is neither square nor round...' and so on, employing thirty-four negations in all. Yet these are not simple negations, but point to the actual reality of the 'self'. The concept of this cosmic life cannot be fully conveyed by a straightforward description. It is expounded here as the 'entity' which can only be expressed in the form of ultimate affirmation through repeated negation.

The Earth, only a small planet, is finite. It is said that even the Sun, that flaming ball of gases which sustains all terrestrial life, will endure only another five billion years. Nevertheless, as we observed earlier in our discussion of Kant's philosophy and Gamow's theory of the universe, many modern sciences, such as astro-nuclear physics and quantum mechanics, are contributing to a greater understanding of the boundless universe. This, I believe, has at last set the stage for a new phase in the contemplation of life.

KIGUCHI: I understand. As an astronomer, my role is to pursue scientific research both intensively and extensively, but at the same time I feel the need to probe further, both intensively and extensively, into the enigma of life itself. On the other hand, as science advances, the understanding of Buddhism will advance all the more rapidly.

IKEDA: That is true. In Buddhism, we often find such terminology as 'three thousand realms in a single life-moment', '*sanzen-jintengō*', '*gohyaku-jintengō*', 'the mutual possession of the Ten Worlds', 'the three truths of temporary existence, non-substantiality and the Middle Way', 'the ten non-dualities or onenesses', 'the twelve-linked chain of dependent origination', and so forth. The clear indication of number and quantity in these concepts reflects the precise approach Buddhism takes in its view of life and the universe. Therefore, I feel it will be most fruitful for 'scientific minds' to study and ponder the Buddhist teachings.

A Buddhist View of Extraterrestrial Life

IKEDA: I believe it was more than twenty years ago that Jōsei Toda, himself a mathematician, said to me, 'Three billion people are now living on Earth, but a hundred years from now, not one of them, myself included, will remain.' He added, 'When I thought about that seriously, it frightened me. And the study and practice of religion – specifically, the supreme teaching of Buddhism – became for me a compelling necessity.' This anxiety about the future was one of his principal motives for conversion.

SHIMURA: Mr. Toda was perceptive. But it seems that most people today, pressed by the demands of their work and other responsibilities, do not individually sense the seriousness of this matter to the same degree.

IKEDA: Human existence may be infinitesimally small in comparison to the vast stretches of the universe. But, for this very reason, I feel all the more determined to strive earnestly toward an understanding of both my own existence and the cosmos.

War – the killing of human beings by other human beings – should never be tolerated. Above all we must abhor atomic and hydrogen bombs, which surely represent the ultimate evil.

In the boundlessness of space, human existence is a rare phenomenon. We must spend our lives striving together to ensure that all humanity can live happily, in peace and without regrets.

In this connection, I recall that Pascal stated in his *Pensées*, 'The eternal silence of infinite space terrifies me.'

KIGUCHI: Three hundred years later, the astronauts who landed on the Moon made comments reminiscent of that passage from *Pensées*.

SHIMURA: Frankly speaking, do you think that somewhere in the universe there may be living beings similar to ourselves?

IKEDA: I think it would not be at all unusual if there were. As I mentioned earlier, life can be conceived of as a reality permeating the universe, in a manner analogous to that of electro-magnetic or gravitational waves. If there are planets somewhere possessing the conditions necessary for life to emerge, then it will surely appear.

With regard to the question of whether life exists elsewhere in the universe, Rector Logunov of Moscow State University told me, 'The possibility exists.' An official of the Ministry of Science in Moscow also said, 'It is possible.' And an American sociologist I met the other day told me, 'I believe it does.' Dr. Toynbee also acknowledged this possibility. Thus, among both scientists and sociologists we find those who agree on the possibility of extraterrestrial life. What do you think, Dr. Kiguchi?

KIGUCHI: If the necessary conditions should be present, it seems inevitable that life would emerge. However, I doubt that we will find any 'E.T.s' living in our own solar system.

SHIMURA: In the immensity of space, the distance from here to the Moon is only slightly more than one light-second, that is, 1.3 seconds travelling at the velocity of light. The distance to Alpha Centauri – the fixed star other than the Sun, closest to the Earth – is 4.3 light-years. Though human beings have travelled to the Moon, the distance covered thereby amounts to less than one hundred millionth of that lying between the Earth and the nearest star outside the solar system. We have journeyed only a very short distance from this planet.

IKEDA: I heard recently from someone that the only man-made structure on Earth visible from the Moon with the naked eye is the Great Wall of China. By the way, Dr. Kiguchi, how far is one light-year in terms of kilometres?

KIGUCHI: As you know, extraordinarily large numbers are popularly called astronomical figures. The figures used in connection with space are so enormous that they bear virtually no relation to our ordinary experience of living. They are truly staggering to our limited, earthbound perceptions. To answer your question, since light travels at nearly 300,000 kilometres per second, one light-year would be about 9.5 trillion kilometres. To illustrate this more concretely, in a single second, light travels a distance equal to about seven and a half times the Earth's circumference.

IKEDA: Buddhism employs many parables to render abstract concepts and difficult doctrines accessible to anyone. These parables explain the essence of the Buddhist teachings in a readily understandable manner, employing concrete objects, situations and events. They are regarded as important Buddhist teachings in their own right. The *Lotus Sutra* itself contains seven parables, including the parables of the three carts and the burning house, the wealthy man and his poor son, and the excellent physician and his sick children.

Shakyamuni's ultimate teaching, in which he expounded a complete view of life and the universe, is contained in the *Lotus Sutra*. The *Lotus Sutra* consists of twenty-eight chapters. Of its four main chapters,[2] the *Juryō* (sixteenth) chapter is the key to the entire sutra.

There are two ways to read the *Juryō* chapter: from the standpoint of *monjō* (literally, 'the surface of the text'), which means to read the literal meaning of the chapter; and from the standpoint of *montei* ('the depths of the text'), which means to read the chapter in light of the supreme Law implicit within it. This supreme Law 'hidden in the depths' of the *Juryō* chapter is without beginning or end. It is immeasurable and boundless, and embodies all truths and virtues. It is the Mystic Law which we believe in.

KIGUCHI: In the beginning of the *Hōben* (second) chapter, another of the four main chapters of the *Lotus Sutra*, Shakyamuni addresses Shariputra, his disciple famed as the 'foremost in wisdom', saying, 'The wisdom of all Buddhas is infinitely profound and immeasurable.' What does 'infinitely profound and immeasurable' mean?

IKEDA: If I may explain it simply, a human life is a microcosm, which, like the external macrocosm, is boundless and unlimited. Its ultimate truth is understood to be 'a reality (*kyō*) as vast as a wide and unfathomable riverbed'. In other words, it has limitless breadth and depth. The Buddha's wisdom (*chi*) completely fills this boundless 'riverbed' that is the reality of all life. If we compare the Buddha's wisdom to water, because its volume cannot be calculated, it is called 'immeasurable'.

Toward an Age of Humility and Contemplation

IKEDA: The *Juryō* chapter is the part of the sutra elucidating the eternity of the Buddha's life. It contains a famous verse section called the *Jigage*. This section begins with the phrase, 'Since I attained Buddhahood' (*Ji ga toku bur-rai*), and ends with the phrase, 'quickly attain Buddhahood' (*soku jōju busshin*). Strangely enough, the first Chinese character of this verse section, *ji*, and the last character, *shin*, when put together, mean 'oneself'. We may understand this to indicate that one's own life is eternally endowed with the Buddha nature.

KIGUCHI: The *Lotus Sutra* is said to contain a complete and perfect view of life and the universe.

IKEDA: Based on the *Lotus Sutra*, we can analyze individual existence in terms of the three properties (*sanjin*), and the universe, in terms of the three truths (*santai*).

SHIMURA: Could you explain the three properties and the three truths?

IKEDA: The three properties are the property of the Law (*hosshin*), the property of wisdom (*hōshin*) and the property of action (*ōjin*). The property of the Law refers to the entity of the Law itself; the property of wisdom indicates profound, penetrating wisdom; and the property of action means actions directed toward establishing the state of freedom.

The three truths are the truths of non-substantiality (*kū*), temporary existence (*ke*) and the Middle Way (*chu*). According to the Nichiren Shōshū doctrine, the *Lotus Sutra* perfectly clarifies the true aspect of the inner realm of life itself. At the same time, it reveals the entire universe to be one integral entity of life, and also clarifies its true aspect and functions. To explain this in simple terms, with the truth of non-substantiality, one can understand the entire universe to be an entity of the ultimate truth. With the truth of temporary existence, one can understand the microcosm of one's own being to be an expression of the ultimate truth. The truth of the Middle Way is the Law that fundamentally integrates non-substantiality and temporary existence. In other words, it is *Nam-myoho-renge-kyo*, the single great truth whose revelation constitutes the purpose of the Buddha's advent. We can interpret the Middle Way as the most fundamental Law, which is embodied in the object of worship.

In other words, the only way that we common mortals can unite the movements of the external world with the inner realm of mind, causing them to function as an integrated whole, is to devote ourselves to the object of worship which embodies the single great truth of the Middle Way.

SHIMURA: In any event, human existence seems infinitesimally small when viewed in contrast with the immensity of the universe or the natural realm. I believe that, for this reason alone, humility is essential as the basis of our thinking.

IKEDA: That is true. We human beings, in our arrogance, have the potential to destroy ourselves and the planet as well.

KIGUCHI: In the latter part of the Edo period (1600-1867), Galileo's heliocentric theory and Newtonian mechanics were introduced to Japan.

SHIMURA: Sakuma Shōzan (1811-1864), a pioneer of Western learning in those days, is said to have read books on astronomy in their original European languages, and even to have made himself a telescope.

IKEDA: I admire the fact that, while contemplating the vastness of the universe, he was also able to consider both Japan and the other nations of the world.

Some say that Shōzan's originality and enlightened thinking stemmed from his research in Western science. Many patriots of the Meiji Restoration[3] flocked to him from throughout the nation, seeking his earnest advice.

KIGUCHI: Somewhat before the end of the Edo period, the theory of Mount Sumeru, which is part of Buddhist cosmology, seems to have excited tremendous controversy.

IKEDA: Yes. However, I believe we may view that controversy as peripheral to the mainstream of Buddhist thought. For example, the world view based on ancient Indian cosmology, which posits Mount Sumeru at the centre of the world, describes only a minor world system. A far greater and in fact boundless immensity is suggested by the expression 'major world system' (*sanzen daisen sekai*), which consists of a thousand minor world systems times a thousand times a thousand. Moreover, the concept of 'dust-particle aeons' (*jintengō*) suggests the limitless flow of time, or eternity. Buddhist texts also employ units of measurement to suggest the idea of 'countless' or 'innumerable' such as a *nayuta* (10^{11}) or an *asogi* (10^{51}). These expressions or analogies contain profound insights into the truth and were developed to express those insights. Therefore, the debate over their literal content itself is not purposeful.

KIGUCHI: Certainly, it would amount to confusing the means with the end. In the realm of science, too, illustrating an abstract theory with a familiar, concrete example not only serves as a powerful weapon of logic but also serves to further the understanding of the scientists themselves. Deepening our comprehension in this way becomes a significant key to the progress of science.

The Meaning of the Ten Worlds

SHIMURA: Incidentally, what we call 'black holes' in space seem to have become a popular topic.

IKEDA: Black holes always remind me of the state of Hell as described in the sutras. At least they serve as a good illustration of that condition. Dr. Kiguchi, would you please explain to us a little about black holes?

KIGUCHI: To put it simply, black holes are said to occur when a star in its final stages collapses inward on itself, generating an intense gravitational force that destroys everything coming within its range. A black hole is said to represent the 'death' of a star.

In his book entitled *The Key to the Universe*, the British astro-nuclear physicist Nigel Calder says, in reference to black holes, 'If the poets were right in saying hell was black the physicists glimpsed it in their theories.' He goes on to illustrate in extremely graphic detail the nature of their gravitational force.

For example, he writes, '... the fate of the imagined space traveller who stumbled upon a black hole became a commonplace way of describing the extraordinary work of gravity, in and around a black hole. Before being trapped and crushed, the unwary astronaut would first be stretched into spaghetti.

'The first hint of trouble might be his hair standing on end, his feet and hands feeling heavy, his head light. The astronaut's blood would drain into his limbs, bringing merciful unconsciousness before the gravity rendered his body into meat, into molecules, into atoms, and eventually into a long beam of particles hurtling into the black hole.' From Einstein's gravitational theory we can accurately deduce the state of a human body being stretched out like spaghetti and torn to shreds under immense gravitational forces. Truly, black holes are like a realm of 'hell' within the physical universe. And, as Mr. Ikeda has said, they serve to illustrate the concept of Hell in Buddhism.

IKEDA: I know little of black holes, but from what I have heard, I gather they are like gravitational 'mouths' scattered throughout the universe, gaping and waiting.

Moreover, should anything fall into them, it would instantly be crushed by their gravitational force and swept away into an oblivion beyond our knowledge to comprehend. It is said that inside a black hole there is no passage of time or expanse of space.

KIGUCHI: That's correct.

IKEDA: It is said that finally the very intensity of their gravitational forces causes them to evaporate and vanish into the universe. In any event, they are an image of hell.

KIGUCHI: According to the theory called the 'Hawking Process', developed by the British astronomer Stephen W. Hawking, black holes do in fact evaporate and disappear.

SHIMURA: Speaking of hell, the Chinese classic *Shih-pa-shih-lueh* (*Abridgement of the Eighteen Histories*) states, 'The depraved appear loyal, and the deceitful seem trustworthy.' Such duplicity seems to be the usual state of those who dwell in the three evil paths of Hell, Hunger and Animality.

IKEDA: Just now we were discussing the state of Hell, but in fact, as Nichiren Daishōnin's *Kanjin no honzon shō* (*The True Object of Worship to Observe One's Own Life*) states, 'Life at each moment is endowed with the Ten Worlds.'

Buddhism classifies life's changing manifestations – that is, the conditions sensed in the depths of one's being – into ten states: Hell, Hunger, Animality, Anger, Humanity, Heaven (or Rapture), Learning, Realization, Bodhisattva and Buddhahood. These are life's specific aspects and directly relate to one's happiness or unhappiness. Moreover, the *Sōkanmon shō* reads, 'Grasses and plants, trees and forests, mountains and rivers, the great earth and a single speck of dust – each is endowed with all Ten Worlds.' Using the analogy of fire latent within flint, the *Honzon shō* explains that the Ten Worlds are inherent in all existences in the universe.

Moreover, the environment wherein living beings of the Ten Worlds dwell also possesses the Ten Worlds within itself. For example, the place where people in the state of Hell live is itself Hell.

KIGUCHI: I've heard that there are many kinds of hell described in the Buddhist sutras.

IKEDA: That's true. According to one explanation, there are eight major hells, each of which has sixteen subsidiary hells, making a total of 136 hells.[4] The eight major hells are: (1) the hell of regeneration, where the bodies of sufferers are slashed with swords but immediately regenerate to undergo the same torment repeatedly; (2) the hell of black ropes, where sufferers are cut and sawed to the measure of black cables; (3) the hell of crushing; (4) the hell of wailing; (5) the hell of great wailing; (6) the hell of burning heat; (7) the hell of great burning heat; and (8) the hell of incessant suffering. Concerning the hell of incessant suffering, the Daishōnin's *Ken hōbō shō* (*A Clarification of Slander*) states, 'Were the Buddha to expound in detail the sufferings of this hell, those listening to him would vomit blood and die. Therefore, the Buddha refrains from giving a detailed description.'

KIGUCHI: While black holes remain an enigma even to scientific researchers, Dr. Calder, as I mentioned before, calls them a glimpse into the blackness of hell, describing in the light of scientific theory how their gravitational force would tear a human body to pieces.

SHIMURA: We often use the phrase 'heart-rending grief'. Recently a medical scientist in the United States reported the case of a patient whose heart ruptured, presumably because of intense grieving, since no other causal factors could be observed.

IKEDA: Perhaps it was indeed due to unbearable anguish. The universe and the human being himself still embody many mysteries yet to be solved.

KIGUCHI: Would you please briefly explain the remaining nine worlds, from Hunger through Buddhahood?

IKEDA: Nichiren Daishōnin clearly grasped the essence of life in the Ten Worlds with his profound insight. His treatise, *Honzon shō*, concisely explains the Ten Worlds where it states, 'When we look from time to time at a person's face, we find him sometimes joyful, sometimes enraged, and sometimes calm. At times greed appears in the person's face, at times foolishness, and at times perversity. Rage is the world of Hell, greed is that of Hunger, foolishness is that of Animality, perversity is that of Anger, joy is that of Rapture, and calmness is that of Humanity.' I feel it is important for science to arrive at an understanding of the true nature of the 'self' that experiences these states.

KIGUCHI: I agree. As you pointed out earlier, science must return to the human being.

SHIMURA: A book entitled *The Body* by the French ethnologist Françoise Loux has recently attracted attention. Considering the body as the centre of the universe, the author interprets it as permeating the cosmos yet grounded in the earth of daily life, and, so to speak, as a bridgework between heaven and earth. His research struck me as quite unique.

IKEDA: I would like to think that the study of the human being is becoming a trend, not only in academic circles, but of the times as well.

Buddhism regards the human being as an entity of *shikishin funi*, or the oneness of physical and mental phenomena. In simple terms, *shiki* corresponds to the body and *shin* to the mind. A human being embodies the essential oneness (*funi*) of these two.

KIGUCHI: So far, however, science, for the most part, has not studied the human being as a totality but rather focused on isolated aspects.

IKEDA: In any event, science and medicine are still groping for the basis of human existence and, in groping, have reached their present stage of development. The French philosopher Henri Bergson and the Swiss psychologist C.G. Jung focused primarily on the mental and spiritual side of the human being.

SHIMURA: Let us discuss this aspect of our topic later on. Coming back to the Ten Worlds: with respect to the world of Hunger, there is a well-known story about Shakyamuni's disciple Maudgalyāyana (Jap Mokuren) and his deceased mother.

Maudgalyāyana's Occult Powers

IKEDA: Maudgalyāyana, famed among Shakyamuni's disciples as the foremost in occult powers, scanned the universe with his divine eyesight and saw his mother suffering in the realm of

Hunger. The *Urabon*[5] *gosho* states, 'In the realm of hungry spirits was his own mother. She had nothing to eat or drink. Her skin was like a plucked fowl's and her bones, like so many rocks. Her head looked like a large globe, her neck was as thin as a thread, and her belly was swollen like the vast ocean. With her mouth open and her hands pressed together begging, she reminded him of a starved leech sucking blood from a man's face.' When Maudgalyāyana tried to send her food by means of his occult powers, the food burst into flames. I believe the flames here may be understood to represent the driving sense of overwhelming impatience coming from within the life of one who has fallen into the state of Hunger.

KIGUCHI: It is a state of life dominated by insatiable desire.

SHIMURA: Our abnormally consumption-orientated society is also characterized by insatiable desire, wouldn't you say?

IKEDA: You are quite right. Moving on to the world of Animality, the *Niike gosho* (*Letter to Niike*) states, 'Animality is to kill or be killed.' The Daishonin also describes this life-tendency in the *Sado gosho* (*Letter from Sado*), where he writes, 'It is the nature of beasts to threaten the weak and fear the strong.' Animality is a state of brute stupidity in which one is totally caught up in instinctual reactions to immediate affairs and makes no attempt to grasp what is most important or essential in life.

SHIMURA: Earlier you quoted the passage where the Daishōnin states that perversity is the world of Anger. What exactly does he mean by perversity?

IKEDA: This expression points to human egotism and conceit. Nichikan Shōnin (1665-1726), the twenty-sixth high priest of Nichiren Shoshu, writes in his *Sanjū hiden shō* (*The Threefold Secret Teaching*), 'Asura is eighty-four thousand *yojana* high, and the waters of the four great seas scarcely lap its knees.' This passage describes the state of someone inflated with pride and under the illusion of his own greatness.

KIGUCHI: The Sanskrit word *asura*, from which 'Anger' (Jap *shura*) derives, originally referred to contentious demons.

IKEDA: One in this state often acts aggressively to camouflage his own cowardice and lack of power.

SHIMURA: Actually, it is an ugly sort of self-exposure.

The Six Paths and the Four Noble Worlds

IKEDA: Hell, Hunger and Animality are together called the three evil paths, and, with the addition of Anger, they are called the four lower worlds. Buddhism therefore clarifies in detail the life-states of the unhappy. The life-condition of Humanity is defined as calmness. We might think of it as the state of one who has established a humane self.

Buddhism postulates eight distinctive qualities of human beings: sagacity, excellence, acute consciousness, sound judgement, superior wisdom, the ability to distinguish truth from falsehood, the ability to attain enlightenment, and good karma from past existences.

KIGUCHI: I see. What about the world of Heaven, or Rapture, as it is also called?

IKEDA: Well, to be brief, as a passage states, 'Joy is Rapture.' According to one explanation, Heaven is divided into a total of twenty-eight realms: six in the world of desire, eighteen in the world of form and four in the world of formlessness. I think we may understand this as indicating the different levels of joy that people experience, from the gratification of instinctual desires to a sense of deep fulfilment in life. According to Nichikan Shōnin, 'Heaven resides in a palace.' This state can also indicate a blessed environment.

However, even those who dwell in Heaven are subject to the five signs of decay. The state of joy experienced in the world of Heaven is momentary, not lasting, and is destined to decline.

The above states are called the *six paths*. According to the

Buddhist teaching, common mortals go back and forth or transmigrate among these six states in lifetime after lifetime. However, human beings are not satisfied with eternal transmigration in the six paths, but strive, through repeated study and effort, to attain the higher life-states of Learning and Realization. More fundamentally, they seek the supreme states of Bodhisattva and Buddhahood, wherein one dedicates himself to humanity and to society, exerting himself for the salvation of the unhappy. This striving to emerge from the six paths and attain the four noble worlds of Learning, Realization, Bodhisattva, and ultimately Buddhahood characterizes the true nature of the 'self' innate in life, I believe. I would like to discuss the four noble worlds in more detail on some other occasion.

Passing the Sun in Eight Minutes

SHIMURA: To change the subject a bit, the other day, a certain painter who also writes interesting essays proposed the following analogy to me. Suppose someone catches the flu and is confined to bed. If this person were magnified to the size of the Earth, then a single influenza virus would proportionately expand to about the size of a football. From this perspective, a person catching a cold would be comparable to the Earth being knocked off balance by a football. What do you think, Dr. Kiguchi, is this analogy valid?

KIGUCHI: Well, a virus constitutes the smallest of the pathogens, ranging in size from about 0.01 to 0.4 micron. Thus, as far as size goes, the comparison is correct. While we cannot see viruses through an ordinary optical microscope, we can actually obtain photographs of them with the help of an electron microscope.

SHIMURA: There is a Dutchman named R. Hawin who collects such extraordinary ideas. Using some of his fascinating data as reference, we may consider the concept of the light-year, which we discussed earlier.
 Suppose that light reaches us tonight from a fixed star in the

Magellanic Clouds. It is, so to speak, as though prehistoric Peking Man who used to burn wood in his cave several hundred thousand years ago were sending us his 'visual greetings'. From this perspective, we can understand that the light from that star has indeed travelled a long, long way.

IKEDA: I've heard that if one were to board a spaceship travelling at a speed of nearly 300,000 kilometres per second and head for the Milky Way, he would pass the Sun in a mere eight minutes.

Forty-five minutes after departure he would pass the vast bulk of Jupiter, and in eighty minutes he would reach Saturn. After five hours, he would arrive at Pluto and then hurtle beyond into the limitless expanse of deep space. After a flight of some four years, Alpha Centauri, the fixed star nearest to the Sun, would come into his view. Six hundred years would pass before he could see Rigel, a blue star of the first magnitude in the constellation of Orion. I remember reading something of this sort. In any event, even moving at the speed of light, space travel requires immensely long stretches of time. The universe is indeed inconceivably vast.

KIGUCHI: That's true. Under these circumstances, several generations of astronauts would be needed to carry on the navigation of a spaceship, passing on the task to their children and grandchildren. Actually, without a rocket that can travel faster than the speed of light, navigation of deep space seems impossible. However, it is by no means impossible in terms of Einstein's special theory of relativity.

IKEDA: You refer to the so-called 'Urashima effect'. It is also known as the 'clock paradox' or 'twin' paradox'. According to this theory, time, as experienced by those travelling in a spaceship, in relation to time on Earth, will slow down as their vessel approaches the speed of light.

KIGUCHI: That's right.

NOTES: CHAPTER II

1 'Danger years' (Jap *yakudoshi*, according to Japanese folk belief, one of the years of bad luck. It is believed that during such a year a person is most likely to experience calamities or misfortune. The ages 25 and 42 for men and 19 and 33 for women are usually considered critical years. Of these, age 42 for men and 33 for women are believed to be especially critical.
2 Four main chapters of the *Lotus Sutra*: according to the T'ien-t'ai doctrine, the *Hōben* (second) and the *Anrakugyō* (fourteenth) chapters are essential to the first half of the *Lotus Sutra* known as the 'theoretical teaching' (*shakumon*), and the *Juryō* (sixteenth) and the *Fumon* (twenty-fifth) chapters, to the latter half known as the 'essential teaching' (*honmon*). These four are called the 'four main chapters' of the *Lotus Sutra*.
3 Meiji Restoration: here it refers to the *coup d'état* of 3 January 1868, in which anti-shogunate forces led by the great southern domains of Satsuma (now Kagoshima Prefecture) and Chōshū (now Yamaguchi Prefecture) seized the Imperial Palace in Kyoto and declared the restoration of power from the Tokugawa shogunate to the Emperor.
4 136 hells: the eight major hells and their sixteen subsidiary hells constitute 136 hells ($8 \times 16 + 8$).
5 *Urabon*: a memorial service to save the deceased from the suffering of hell. This ceremony originates from the story of Maudgalyāyana.

III
Investigating the Mysterious Laws of the Universe

'The finity reveals the infinity; an instant contains eternity.' How does intuitive Buddhist wisdom grasp the time and space of the great universe?

'The Earth Is Blue and Very Beautiful'

KIGUCHI: Considering our orientation as we approach the twenty-first century, it will make me very happy if people can come to understand something about astronomy.

SHIMURA: It may be that our discussion is sometimes difficult to understand. The technical vocabulary of astronomy is certainly challenging, but that of Buddhism may in some respects seem quite beyond the comprehension of the lay person. I suppose this is inevitable. Consider, for example, the bewildering multiplicity of names for the Buddha. He is not only called the Buddha, but also the Tathāgata or Thus-Come One (Jap *nyorai*), as well as the Bhagavat or World-Honoured One (*seson*).

IKEDA: That's true. Depending on the sense in which he is referred to, the Buddha is known by many names. In addition to those you have already mentioned, we find that he is called 'he who can endure' (*nōnin*), 'person worthy of offerings' (*ōgu*), 'possessor of right and universal knowledge' (*tōshōkaku*), and so forth.

KIGUCHI: In the realm of science, terms such as those used for the names of elements and molecules are standardized throughout.

IKEDA: In Buddhism, a single truth or experience may be interpreted from various levels or perspectives. For example, we may discern a surface or literal meaning of something (*ichiō*) and also recognize its significance from a deeper perspective (*saiō*). We may view truth from the standpoint of both the general (*sō*) and the specific (*betsu*), or from both an immediate perspective (*tōbun*) and a more thorough-going one (*kasetsu*). The *Lotus Sutra* can be read in terms of its literal meaning, as the teaching of the historical Buddha, Shakyamuni (*monjō*), or in terms of the meaning hidden in its depths, in light of Nichiren Daishōnin's advent (*montei*). Moreover, we can approach the truth as revealed in the text of the sutras (*kyōsō*) and as observed within the mind (*kanjin*). Buddhism is an immensely profound teaching that must be grasped in its multiple dimensions to be truly understood.

KIGUCHI: From what you have just said, I feel we can identify an essential difference in the respective approaches to truth of a basically materially-oriented civilisation and a more spiritually-oriented one.

SHIMURA: Incidentally, I hear that a total eclipse of the Sun will occur on 11 June 1983.

KIGUCHI: That's right. It will be observable from all over the world including the North and South poles, although there are some places where observation may prove difficult. It is said that this will be one of the most readily visible solar eclipses of the century. As you know, a solar eclipse occurs when the moon passes between the Earth and the Sun, blocking all or part of the light reaching the Earth directly from the Sun.

IKEDA: Solar eclipses have been regarded as 'bad omens' since ancient times. However, with the advance of human knowledge, many of our notions about the significance of natural phenomena have changed a great deal in the course of human history. The expansion of knowledge and the progress of science have helped to banish many irrational fears and superstitious beliefs.

Ancient peoples were often threatened by the forces of nature. So they prayed to the gods of wind, water and fire to soothe their tempestuousness and hoped thereby to avert calamity. Such prayers demonstrate a certain primitive faith or religious sense.

However, no matter how far knowledge or science may advance, it cannot resolve the problem of suffering inherent in human life, or change negative karma. Indeed, we say that such sufferings have become all the more complex and serious despite the advance of science. Consequently, we need a higher religion, one which incorporates a universal and eternal view of life and the cosmos. We may say that the depth of a religion is in part determined by the clarity of its views on life and the universe, and by how well it substantiates its views theoretically.

Outer Space: Its Colour, Winds and Temperature

SHIMURA: I would like to return briefly to the issue of space travel. In considering the vastness of space, it seems absurd to think of living beings only in terms of the framework of our own solar system. I'd like to ask about our Galaxy – a realm of shining stars measuring some one hundred thousand light-years across. Do you think that in the future it will be possible to travel to the centre of the Milky Way and back within the span of one lifetime? I think some discussion of this point might prove interesting.

IKEDA: Far away in the vast universe lie stars whose light takes an unimaginable number of millions or billions of years to reach the Earth. It seems that the world's scientists are investigating plans for human space travel. In Buddhism we find the doctrine that one moment encompasses eternity, or that eternity is contained within a single moment – a view which completely envelops the vastness of cosmic time and universal space. I wonder if someday it will be possible, through scientific means, to conquer the billions of light-years that space travel would presently require, reducing it to a mere twenty or thirty years.

KIGUCHI: I find Buddhist principles such as the one you just mentioned fascinating, especially in the light of the recent manned space voyage to the Moon. Speaking in terms of the rapid rate of scientific advancement and our present-day calculation, travel through the Galaxy is theoretically possible. Yet I wonder if it will in fact ever actually take place.

IKEDA: It seems that for the present, we are limited to travel within the solar system, and that journeying to more distant stars is not, as yet, feasible. By the way, what colour is outer space?

KIGUCHI: I understand that it varies considerably.

IKEDA: What colour would it be, for instance, about halfway between the Earth and the Moon?

KIGUCHI: Well, the stars shine white, red and yellow in pitch darkness. The bright rim of the Earth would appear like a crescent moon, with the rest of its surface a dull red.

IKEDA: And if one were to go farther?

KIGUCHI: The sky would be pitch black, but one could see groups of stars and the Sun shining brightly.

IKEDA: Is there any wind? Or is outer space utterly still?

KIGUCHI: Winds blow everywhere in space. The presence of a solar wind has been confirmed, and, in theory at least, the existence of galactic winds has also been postulated.

IKEDA: If one were momentarily to leave his spaceship and inhale 'space', would it taste more delicious than our own air?

KIGUCHI: I think it would probably depend on the local density of cosmic dust. Since the greater part of space consists of helium and hydrogen, I don't suppose it has any particular taste or odour. However, there are various organic molecules present in the gaseous clouds among the stars that might have a perceptible taste. Of course, so far no one has actually tasted it.

IKEDA: An astronaut exclaimed, 'The Earth is blue and very beautiful.' What about temperature in space?

KIGUCHI: It varies according to where in space you are, but I imagine that it grows colder the farther away you go from the Sun.

IKEDA: What do you think about the probability of life emerging elsewhere in the universe?

KIGUCHI: In outer space itself, the probability is close to zero, but there must be, throughout the universe, a great number of planets with conditions similar to the Earth's. Scientists conjecture that life may well have emerged on such planets.

IKEDA: Then we may imagine that somewhere in the universe there exist civilizations similar to our own.

Life Is More Precious than the Earth

SHIMURA: Are the events of the movie *E.T.* really conceivable?

KIGUCHI: No, I hardly think so. Only in the movies could an extraterrestrial being land on Earth and befriend a child.

SHIMURA: I hear that the movie was based on the Search for Extraterrestrial Intelligence (SETI) activities being conducted in the United States.

IKEDA: I believe such an encounter is strictly fictional and could never happen in reality. But people feel a strange attraction to the unknown. They long to scale unclimbed mountains, or yearn for new discoveries, such as those made by Columbus or for adventures like those of Marco Polo. Amid the reality of modern society, which so many people find oppressive, it is only natural that they turn their eyes toward the universe.

Quite apart from this longing for the cosmos, Buddhism adopts as a premise the existence of numerous worlds throughout the universe inhabited by living beings.

KIGUCHI: I feel that this is an important basic premise to accept. It is valuable for me as a scientist, and I'd like to ask you about it at length.

SHIMURA: These days, even while we speak of 'the universe', we sense a growing apprehension that outer space – a neutral realm above all differences in human society that politically divide the people of the Earth – might in the future become another battlefield.

KIGUCHI: That's why, as a researcher, I've come to keenly feel the necessity to understand the cosmos more deeply in terms of its relationship to life.

SHIMURA: In other words, it is vital that we recognize it as 'life space', wouldn't you say?

IKEDA: That's right. Space must never become an arena for war.

SHIMURA: The philosophy of Buddhism teaches reverence for life and for the environment that supports it.

IKEDA: A single human life is more precious than the Earth. Buddhism is the teaching that thoroughly elucidates the laws of life. I wish to assert that the only way to fundamentally ensure the survival and to protect the peace of the world and the universe is through a deep understanding of Buddhism, both in its theoretical aspect and in terms of its correct application to one's individual life and to society. This is what I believe.

KIGUCHI: We may say that our times have reached the point where there is great need for the establishment of a 'theory of peace' of a high dimension, one that can consistently comprehend the universe as well as the Earth itself.

How Vast Is the Cosmos?

SHIMURA: I've devised a method by which one can gain some sort of real sense of the boundlessness of the universe. For example, when our present discussion is printed, commas will be inserted in the text, say, to indicate pauses in speech. For the sake of discussion, let one comma correspond to the size of the Earth.

IKEDA: That's about 0.5 millimetre in diameter. Fine, we'll make that our gauge for considering the other heavenly bodies. Dr. Kiguchi, I realize this is on the spur of the moment, but how far would the distance then be to the Moon, the nearest heavenly body?

KIGUCHI: Well, roughly it would correspond to the thickness of a forefinger. About fifteen millimetres, I'd say.

IKEDA: And to the Sun?

KIGUCHI: About six metres. Say, the length of a limousine.

SHIMURA: This is beginning to sound like a quiz show, but how far would it be to the closest fixed star other than the Sun?

KIGUCHI: Well, the distance from Tokyo to Kyoto on the bullet train measures five hundred kilometres. Let's say about three times as far as that.

SHIMURA: Then, to return to our original discussion, how far would it be from the Earth to the centre of the Milky Way, measured in terms of 'commas'?

KIGUCHI: Well, let me see. The distance to the centre of the Galaxy is about two billion times the distance between the Earth and the Sun, a distance utterly beyond our ability to grasp.

IKEDA: To make it a bit simpler to understand, if we scale down the Earth to the size of a comma, then the centre of the

Galaxy would be shining in the distance some twelve million kilometres away.

The other day, while talking to several people in Niigata, this *haiku* by Bashō (1644-1694) suddenly sprang to mind:

> The rough sea:
> Arching over Sado Island –
> the Milky Way.

SHIMURA: Bashō was a unique Japanese artist who integrated his observations of the universe, human affairs and all living creatures in his own profound introspection. Incidentally, we seem to be enjoying quite a *haiku* boom lately.

IKEDA: We are, indeed. It seems to me that Bashō strove to grasp the entire universe within his mind. Similarly, I believe that the recent popularity of *haiku* has stemmed from people's attempts to confirm their own inner world by capturing objects before them in verse.

KIGUCHI: Life in the modern world is terribly hemmed in by superficiality.

SHIMURA: Speaking of poetic feeling reminds me of the writer Yasushi Inoue, who, after reading a collection of Mr. Ikeda's poems, entitled *Seinen no fu* (*Ode to Youth*), was deeply moved and remarked that, in particular, the poems 'The Universe' and 'Mother' could not have been written by any so-called poet in the world today. I recall that Mr. Inoue, himself a poet, said that Mr. Ikeda encompasses in a single moment within his heart the deep meanings of the vast heavens and human affairs. He also praised the decision to honour Mr. Ikeda with the title of poet laureate.[1]

IKEDA: I was greatly flattered. In any event, no matter how much we may be caught up in the press of our daily affairs, because life is limited, we should make room in our lives to cherish the poetic spirit, which is as fresh as morning air. Similarly, I wish to live my life with a heart that is always pure and clear. Nichiren Daishōnin teaches that our lives are 'the palace of the ninth consciousness, the unchanging reality which reigns over all life's functions'.

By the way, how far would it be to the Andromeda Nebula, the closest neighbour of our Galaxy? I realize these calculations must be difficult, but after all, Dr. Kiguchi, you are a professional (laughter).

KIGUCHI: That nebula can be seen only from the northern hemisphere. Judging from available data, I would say it lies about 1.9 million light-years away, about sixty times the incredible distance from the Earth to the centre of the Milky Way that was mentioned before.

SHIMURA: How many stars are there in our Galaxy? There seem to be so many that they form a solid glittering path, as the name 'Milky Way' connotes.

KIGUCHI: There are said to be about one hundred billion stars in the Milky Way, so we might well say that it contains countless stars.

IKEDA: Countless indeed. I believe I read somewhere that there are enough stars in the Galaxy to provide thirty solar systems for every single human being on Earth, from newborn babies to grandfathers and grandmothers. Of course, not everyone would want them (laughter). The universe is truly immeasurable, and the Earth no more than the tiniest speck.

KIGUCHI: We will be discussing this subject later on, but some scientists estimate that in the Milky Way there may be ten million planets where civilizations could exist.

IKEDA: The human life span is less than a hundred years, but it takes 1.9 million light-years, in terms of Earth time, to travel to the next nebula. In the light of this, one couldn't go there even if one wanted to. However, I understand that time slows down as one travels through space.

The Tale of Urashima Tarō

KIGUCHI: Yes, it's true that time slows down as one's speed approaches the speed of light. This is the so-called 'Urashima effect', or 'twin paradox'.

SHIMURA: According to the folktale,[2] the three years Urashima Tarō spent in the sea-god's palace corresponded to seven hundred years in the present world. I don't know what basis was used for calculation.

IKEDA: This legend is mentioned in such texts as the *Nihon shoki* (*Chronicles of Japan*) and the *Man'yōshū* (*Collection of Ten Thousand Leaves*),[3] compiled in the eighth century. It is a tale which expresses man's desire for lasting youth and long life.

SHIMURA: In a different version of the story, Urashima Tarō goes, not to the sea-god's palace but to Mount P'eng-lai,[4] and in others, the hero is not Urashima Tarō at all but a boy who is really the spirit of some constellation, such as Subaru (Pleiades) or Amefuriboshi (Hyades).

KIGUCHI: That's interesting, although it doesn't seem to have any particular relation to Buddhism.

IKEDA: Similar folktales are found in many parts of Asia, so some of them must have been introduced to Japan in ancient times. In the letter Nichiren Daishōnin addressed to a nun disciple, Kōnichi-bō, he says that on hastily opening a letter from her which proved to contain sad news, he experienced disappointment such as that which Urashima must have felt upon opening his casket.

SHIMURA: In the Edo period (1600-1867), the great playwright Chikamatsu Monzaemon wrote *Urashima nendaiki* (*Chronology of Urashima*), and in the Meiji period (1868-1912), we find such works as Shōyō Tsubouchi's *Shinkyoku urashima* (*Urashima to a New Tune*), Rohan Kōda's *Shin urashima* (*The New Urashima*), and Ōgai Mori's *Tamakushige futari urashima*

(*The Two Urashimas*). All of them seem to have been quite popular.

KIGUCHI: But if you tell the story of Urashima Tarō to children today, they'll protest that it's just a fairy tale.

IKEDA: These days children probably find the 'Urashima effect' theory of space-time more attractive than the tale of Urashima Tarō itself. Dr. Kiguchi, would you please explain this relationship in simple terms?

KIGUCHI: Well, very simply, we are talking about a concept developed by Einstein on the basis of Galileo's principle of relativity and Newtonian mechanics. In brief, certain physical laws operate within our Galaxy. The very same laws are operative within other galaxies as well. However, because there is a distortion or curvature in space between any two galaxies, when one is observed from the standpoint of the other, it appears as though completely different laws apply there. These laws or principles, which appear upon direct observation to differ, may be compared by means of their alteration in the scale of time-space. Moreover, Einstein discovered that these alterations can be calculated.

IKEDA: In short, our experience of time and space as we perceive it would appear to be quite different if viewed from heavenly bodies.

KIGUCHI: Space is warped or distorted everywhere. The space surrounding the Earth is decidedly curved. It was Newton who first surmised that the space around the Earth is where its gravitational field forms and the workings of gravity arise.

Warping of Space and Alterations in Time

IKEDA: Newton's greatness may lie in his ability to have speculated on why apples fall while the stars in the heavens do not.

Some years ago, when I met with Dr. Toynbee, I visited the University of Cambridge. On that occasion, some members of the university staff guided me to a park near the campus, where Newton is said to have seen an apple fall and had his famous insight. People like Newton who awaken to some great principle through their own perception of phenomena are like the *pratyekabuddhas* or men who realize a truth by observing natural phenomena, as is described in Buddhism.

KIGUCHI: The curvature of space can perhaps be best understood in terms of the character of light. Light travels in a straight line, yet we can see stars positioned behind the Sun, which by rights we shouldn't be able to see. We can actually see stars 'hidden' behind the Sun. This is because the warp in space produced by the Sun's gravity deflects the starlight and alters its rate of travel.

IKEDA: In other words, you mean the light of these stars follows the curvature of the Sun in reaching the Earth?

KIGUCHI: That's right. This demonstrates that space is curved. And where space is curved, the flow of time is also altered. Einstein demonstrated theoretically that because space is warped or curved everywhere in the universe, the passage of time is also not constant anywhere.

IKEDA: And the greater the curvature of space, the slower the passage of time.

KIGUCHI: That's right. The curvature of space around the Sun is greater than that around the Earth, so time passes more slowly with respect to the Sun than it does with respect to the Earth.

SHIMURA: Then the sea-god's palace must have been somewhere in space, rather than in the ocean (laughter).

Black Holes without Time

IKEDA: Black holes are areas of the most extreme distortion. At their centre, it is said, time and space completely cease to exist. Inside a black hole, the hands of a watch would not advance but would remain absolutely still.

KIGUCHI: Theoretically that's true. As dying stars condense, their gravitational pull becomes increasingly stronger. Black holes represent their final, collapsed state. As we mentioned earlier, when we compared black holes to the state of Hell, ordinary physical laws are suspended inside them.

SHIMURA: Since gravity results from a warping of space, the interior of a black hole must present an extremely strong gravitational field.

KIGUCHI: The surface of a black hole, which cannot be seen, is called the 'event horizon'. The closer one approaches this horizon, the more the passage of time slows. Near the horizon, even an instant, the merest fraction of a second, would seem a virtual eternity to us.

IKEDA: Once one were about to enter a black hole, time would stop for him. This is indeed analogous to the state described in Buddhist texts as 'the great citadel of the hell of incessant suffering'. Nichiren Daishōnin states, 'Life at each moment encompasses both body and spirit, and both self and environment of all sentient beings in every condition of life, as well as insentient beings – plants, sky and earth, on down to the most minute particles of dust.' Here, 'in every condition of life' refers to the Ten Worlds. Black holes truly manifest a hellish aspect of the insentient realm of space. They suggest the experience of time in the life-state called 'the hell of incessant suffering', wherein one's anguish seems to last an eternity.

KIGUCHI: Moreover, inside the gravitational field of a black hole, a single moment of physical time would pass at different rates with respect to the top of one's head and the soles of his feet. To illustrate, suppose that an infant's head were in an instant to become that of an old man, while his feet were to remain those of a child. Such distortions of time are indeed hellish. Black holes, from which one could never escape, might well be called the 'hell' of the cosmos. The passage of time as one would experience it there, like the passage of time as perceived in the state of Hell, would seem unbearably slow.

SHIMURA: The opposite principle is illustrated by the story of Urashima Tarō. The period he spent in the sea-god's palace was like time experienced in the joyous, unfettered state of Rapture or Heaven. Although it seemed to him to have lasted three years, it was in reality seven hundred. I think the legend may be seriously considered in this light. If it were no more than a fantastic fairy tale, it wouldn't have been handed down for more than a thousand years.

IKEDA: Good times pass quickly and inevitably come to an end. The same is true of our human existence. Youth does not last forever, and everything that has form must eventually undergo destruction. All that lives must one day perish. This holds true of all phenomena in the universe.

Buddhism: The Strict Principle of Reform

SHIMURA: The recent phenomenon of transplanting artificial human hearts has caused people to reconsider the problem of death. It is said that in the case of one such patient, Dr. Barney Clark, his artificial heart went on functioning mechanically long after all the functions of his nervous system had completely ceased.

KIGUCHI: I've also heard that while the rest of the body grows older, an area that has been reshaped by cosmetic surgery may remain youthful in appearance. I imagine it can be quite embarrassing.

IKEDA: I believe we have entered an age when people will have to consider science, politics and various other fields in conjunction with the problem of death. In any event, Buddhism grasps the eternal cycle of birth, old age, sickness and death – or formation, continuance, decline and disintegration – as 'the impermanence of all phenomena' or 'the law of emergence and disappearance'. However, rather than focusing on transient phenomena, Buddhism clarifies the eternal, unchanging principle of life which permeates them.

KIGUCHI: In other words, it teaches that if one clings to ever-changing, mutable phenomena, he becomes a slave to desires and illusions, is that right?

IKEDA: That's correct. However, Buddhism also recognizes desire to be a natural expression of life. It teaches that, by awakening to the Law which is eternal and unchanging, one can attain the state in which earthly desires are enlightenment (*bonnō soku bodai*). This seemingly contradictory principle is based on the *Lotus Sutra*. Earthly desires are like poison, and enlightenment, like medicine. These two, one opposing the other, exist within our lives simultaneously. But while one is manifest, the other remains dormant. The reason we can say that earthly desires *are* enlightenment is because when we cause the cosmic life force or the eternal Law to function within our lives, we gain the strength to use earthly desires as a means to manifest our inherent enlightenment. This is why the *Ongi kuden* states that when one burns the firewood of earthly desires, the fire of enlightenment glows before our eyes.

SHIMURA: Are we correct in saying that Buddhism stresses the impermanence of all phenomena precisely in order to reveal the eternal, unchanging Law?

IKEDA: That's right. Those who fail to grasp why Buddhism expounds the truth of impermanence sometimes tend to see only this aspect and come to view Buddhism as a teaching of resignation or nihilism. We can say that such an understanding is incorrect. Buddhism strictly appraises the realities of the world, and, from this standpoint, unfolds the logic of reformation.

SHIMURA: I understand. By the way, one by-product of our pioneering space studies is the rapid development of new weapons systems.

Recognizing Science's Diabolical Side

IKEDA: I understand that nuclear weapons were developed on the basis of insights gained through study of the stars.

KIGUCHI: That's true. As Dr. Victor F. Weisskopf, former head of the American Physical Society, pointed out in his anti-nuclear weapons appeal issued in 1983, nuclear energy is a fundamental energy of the cosmos. It is human beings who have misused it.

SHIMURA: For example, it is said that scientists studying and calculating the process by which stars collapse under the tremendous pull of their own gravity and then finally explode, developed, on the basis of these studies, the theoretical principles for the production of weapons that could destroy the Earth. It is rightly said that space studies hold the potential to radically alter human destiny.

Among those engaged in developing the atomic bomb, with its vast destructive powers, were the American physicists J. Robert Oppenheimer and Edward Teller.

KIGUCHI: These two were pitted against one another at the loyalty hearing of 1953, Oppenheimer opposing the production of the hydrogen bomb. This may have been the first incident of its kind arising from a scientist's own clear perception of the diabolic nature inherent in science.

IKEDA: I always think that Buddhism can provide people with boundless wisdom, not only to perceive the diabolical potential present in science, but to direct science in ways that completely serve the cause of peace. In this connection, we may say that the concrete actions for peace on the part of those who believe in Buddhism reflect an active humanism grounded in daily life.

KIGUCHI: I understand very well. What you have just pointed out is a major issue for scientists. It is a very fine line that determines whether the fruits of one's research will be used for war or made to serve the cause of peace. As an extension of this problem, the scientist's own feeling and convictions become extremely important.

A Journey through Space by the 'Twin Paradox'

IKEDA: Coming back to the subject of the effects of space curvature on the passage of time, it would be interesting to postulate a great space voyage conducted by means of the Urashima effect, or the twins paradox, calculating the slowing of time.

SHIMURA: As our hypothetical destination, I recommend the Andromeda Nebula, 1.9 million light-years from the Earth.

KIGUCHI: The Andromeda Nebula is our neighbour, so to speak. If we travel that far, the Urashima effect will be clearly evident.

SHIMURA: Actually, it seems that scientists do speculate among themselves about such space voyages on a theoretical basis. The projected spaceships used to travel back and forth on such journeys are called *interstellar ramjets*.

KIGUCHI: While in flight, they take in hydrogen gas from space as fuel. Since there is on the average only about one hydrogen atom in every ten cubic centimetres in deep space, the internal structure of these ships which enables them to draw in hydrogen while in flight must be quite amazing. The American scientist Dr. Robert W. Bussard has already announced a tentative plan.

IKEDA: As the spaceship gathers speed after take-off and steadily approaches the speed of light, the passage of time within the spaceship will slow down accordingly. However, those on board will not sense any difference. And, even though the passage of time slows radically within the ship, Earth time will remain the same, so the passengers will undergo the Urashima effect.

KIGUCHI: Inside the spaceship, if the force of gravity that we experience on Earth (9.8 metres/second^2) is constantly accelerated, the passengers will not have any sensation of weightlessness. Rather, they will continue to feel just as they did on Earth.

SHIMURA: According to Einstein's special theory of relativity, no matter how much a spaceship may accelerate, it can never exceed the speed of light.

IKEDA: That's true. The longer we travel through space, the more marked the specific time differential between the Earth and the spaceship becomes. A spaceship could not journey to a close destination such as the Moon, for example, at anything approaching the speed of light, because it wouldn't have enough time to accelerate.

KIGUCHI: For example, en route to the Andromeda Nebula is the Great Nebula of Orion, which is still within our own Galaxy. The starlight from this nebula takes fifteen hundred years to reach the Earth. However, calculating in terms of the Urashima effect, spaceships could traverse this distance in about eight years.

The Buddhist Concept of Subjective Time

KIGUCHI: I was surprised to find that Mr. Ikeda's poem, 'The Universe', which was mentioned before, expresses, accurately yet in beautiful words, laws operating throughout the vast reaches of the cosmos. As someone who had chosen space as the subject of his lifelong study, I learned a great deal from it. For example, it reads:

> The moments of cause and effect are seen simultaneously,
> the limited is pregnant with limitlessness,
> the instant embraces eternity, ...

I feel that in these three short lines, Mr. Ikeda, you have managed to express fully some vital principles of the universe.

SHIMURA: On re-reading 'The Universe', I myself felt that this flash of poetic feeling must have come from Mr. Ikeda's intuitive insight as one who has exerted himself in the practice of Buddhism. How does Buddhism approach the concept of time?

IKEDA: Time in Buddhism is not calculated on a theoretical basis. Rather it is grasped subjectively, as time experienced in the depths of life. That is, it is time as understood by the Buddha's unrivalled wisdom, based on the way that people actually experience it.

Nichiren Daishōnin states in the *Sōkanmon sho*, 'Although past, future and present are divided into three, they are without distinction in the truth of the life-moment.' In other words, both the past and the future can be sensed in a single moment of life – they are encompassed by the present moment.

In Buddhism, past, present and future as mentioned above are called 'the three existences of life'. Sometimes they are also called, respectively: the already-come, the thus-come, and the yet-to-come. 'Thus-come' (Skt *tathāgata*, Jap *nyorai*), as you know, is also a title of the Buddha, and indicates the ultimate reality or truth which manifests itself at each moment as the emergence and disappearance of all phenomena. One who has grasped this reality or truth is called a Buddha. Moreover, the word Buddhism in Japanese is *buppō*, literally, 'the Buddha's Law'. The character for Law here is written with two elements, meaning 'water' and 'to depart'. Water does not remain, but comes and flows on without ceasing. It is said that this character thus symbolizes the ultimate truth expressing itself moment by moment in the continuous flow of the emergence and disappearance of all phenomena.

Eternity in a Moment

KIGUCHI: Then the present does not exist apart from the past and future. Rather, both past and future are to be found within one's life at the present moment. Is this understanding correct?

IKEDA: Essentially, yes. An eternity of time is fused in oneness with the present moment. Thus we sometimes say that 'eternity exists in the moment.' The divisions of past, present and future are created by the flow of consciousness, but in essence, these three are 'without distinction', as the passage I just cited states.

KIGUCHI: In physics, too, the idea of the present and the future, or the past and the present, being completely equivalent, forms the starting point for many theories. One example of this is the principle of the conservation of energy. In many cases, the theories of modern physics can no longer admit perceptions based on the flow of consciousness, so we cannot draw binding distinctions between past, present and future.

SHIMURA: Mr. Ikeda, what is the meaning of 'the truth of the life-moment' in the passage you cited earlier?

IKEDA: Simply stated, the notion of the flow of time results from our consciousness and actions. 'The truth of the life-moment' generally refers to the realm of our 'inner cosmos', and may also be understood as the life of the outer cosmos. The principle of life is inherent and eternal; it is an unchanging truth. However, by means of our life-activities, it manifests as the changing phenomena of past, present and future. Moreover, concerning the relationship between life and time, the Daishōnin speaks of 'the mysterious single Law that simultaneously contains both cause and effect and is called *Myōhō-renge*' in the *Tōtaigi shō* (*The Entity of the Mystic Law*).

KIGUCHI: Generally speaking, even in terms of the most trifling phenomena, cause and effect do not occur at the same time. What is meant by the 'simultaneity of cause and effect'?

IKEDA: Well, to explain this may become a bit difficult, I'm afraid, but as we have said, a single life-moment encompasses the vast reaches of eternity, or we could say that life's eternity is condensed in the life-moment. Accordingly, all causes and their effects are simultaneously contained within the life-moment. But in particular, the 'simultaneity of cause and effect' refers to the 'mysterious single Law'. This is the mystic principle that beings of the nine worlds, or the nine life-conditions of the common mortal (cause), simultaneously possess within themselves the power to manifest the state of Buddhahood (effect).

The Oneness of Beginningless Time and the Latter Day

SHIMURA: How would you explain the principle of *kuon soku mappō* (the oneness of beginningless time and the Latter Day of the Law) in terms of the 'mysterious single Law' – that is, the Mystic Law?

IKEDA: This, too, is rather difficult to explain. Briefly, the word *kuon* means 'the extremely remote past'. In the principle that you have mentioned it indicates *kuon ganjo*, or time without beginning, while the Latter Day of the Law indicates the present time. The word *soku*, which links the two, means 'just as it is', or 'essentially one'. In other words, on the basis of the Mystic Law, timelessness manifests itself, just as it is, in the present moment.

The word *kuon* itself simply means an incredibly vast length of time. It can be interpreted on several levels: in terms of the theoretical teaching of the *Lotus Sutra* (*shakumon*), the essential teaching (*honmon*), the *Lotus Sutra* viewed as Shakyamuni's teaching (*monjō*) and the *Lotus Sutra* as viewed in light of Nichiren Daishōnin's enlightenment (*montei*). However, this subject becomes excessively technical, so I will not go into it here.

Nichiren Daishōnin states in the *Ongi kuden* that '*kuon* means one uncreated, unadorned, just as it originally is.'

We moderns tend to try to grasp everything, including the notion of *kuon*, solely within the limits of time and space. I feel that this may be why we have difficulty understanding how the infinity called *kuon* and the present moment can be 'one'. Let's try a change of approach. As is clear from the passage of the *Ongi kuden*, which I just cited, the true meaning of *kuon* transcends the dimension of time to reveal the Law or ultimate truth of life and the universe – the supreme principle from which all things originate and to which they all return.

Accordingly, the 'oneness of *kuon ganjo* and the Latter Day of the Law', in brief, is not merely a theory of time. It involves the mystic truth that has always been inherent within our own lives and the universe. This mystic truth is encompassed by our lives in the present moment, and, in fact, it is only through the present moment that we can awaken to it.

KIGUCHI: I see. I had thought that *kuon* only meant the remote past.

IKEDA: With respect to its fundamental meaning, we find expressions in Buddhist writings such as 'the Mystic Law of beginningless time inherent in the life-moment' and 'the quality of cause and effect inherent in a single life-moment'. Accordingly, time without beginning or *kuon ganjo* is the Mystic Law; past, present and future are all expressions of the Mystic Law. The mystic truth is the Law of beginningless time; it is *Nam-myōhō-renge-kyō*. Because of the principle that cause and effect are inherent in a single life-moment, rather than a past or a future, it is the present moment which assumes the utmost importance. When we who in the present age of the Latter Day of the Law are basing our lives on *Nam-myōhō-renge-kyō*, the Law of beginningless time, awaken to that Law even partially, then at that moment we manifest within our own lives the oneness of *kuon ganjo* and the Latter Day of the Law. I believe, therefore, that we can grasp this principle as a firm reality only through our practice of faith, that is, our devotion to the Mystic Law.

SHIMURA: When we consider the time-space alterations produced by the Urashima effect, in comparing the passage of time inside a spaceship with that on Earth, we can, from a theoretical perspective, understand that the present moment does indeed encompass both past and future.

KIGUCHI: I believe we may say so. It takes fifteen hundred years to reach the Great Nebula of Orion measured by Earth time, but a spaceship could hypothetically make the journey in a mere eight years. The farther one journeys into the immensity of space, the more the time differential increases with respect to the Earth.

IKEDA: A spaceship could make a round trip to the Orion Nebula and back in sixteen years, but by the time it returned, three thousand years would have passed on Earth. Thus we may say that three thousand Earth years equal sixteen years of the spaceship's time.

KIGUCHI: That's right. Even according to the concepts of physics, if the infinite time-space of the universe is taken as our laboratory, abstruse laws and principles will become very clear.

IKEDA: The ultimate principle of life is completely endowed and perfect; it lacks nothing. It is an absolute principle. Einstein, in formulating his special and general theories of relativity, plucked the strings of his fine intellect, so to speak, producing an exquisite harmony that resonated with the life of the universe.

SHIMURA: The Andromeda Nebula in the next galaxy – the final destination of our hypothetical voyage – is some 1.9 million light-years away. How long would it take to travel there, allowing for the Urashima effect?

KIGUCHI: At one stroke, the time involved would be shortened to a fraction of one 127,000th. The voyage would take a mere fifteen years.

IKEDA: This indeed suggests the principle of the oneness of beginningless time and the Latter Day of the Law, or that of an instant encompassing eternity. The hundreds of billions of stars in the Andromeda Nebula will welcome us with an awesome display of light.

KIGUCHI: Assuming our spaceship returned safely, 3.8 million years would have passed on Earth since the time we left.

SHIMURA: That makes the few centuries that Urashima Tarō spent in the sea-god's palace look like nothing.

IKEDA: That is a delightful fantasy. So that's how much time will have elapsed? I hope that the human race will not have already passed into extinction by the time we returned.

SHIMURA: In the movie *Planet of the Apes*, a spaceship leaves the Earth, journeys out into the vastness of space, and eventually arrives at a certain planet. At the end of the story, the remaining members of the crew learn that the planet is in fact the Earth. Human civilization had degenerated to the level of prehistoric barbarism and the planet had come to be controlled by a society of apes.

IKEDA: We must not allow anything to block the road to survival and to peace.

Life's Mysterious Origins

SHIMURA: To change the subject a bit, toward the end of 1982 I contacted Dr. Carl Sagan in the hopes of discussing with him the topic of possible communication with extraterrestrial intelligence. On the phone, he told me that the subject was of great interest to him. But, as his wife was expecting their baby very soon, he continued, 'I shouldn't leave Ann's side even for a moment. I want to welcome with respect this "messenger from space" who has made a long, solitary journey from a distant cosmos.' From anyone else it might have sounded a bit affected, but I was impressed by these words from a man who has spent so much time and energy trying to penetrate the mysteries of the universe.

IKEDA: So he was engaged in a most enjoyable form of research and speculation in attempting to clarify the nature of life in the universe (laughter).

KIGUCHI: Scientists engaged in space studies are generally romantically inclined anyway.

IKEDA: As we ponder and investigate the subject of whether or not living creatures similar to human beings exist somewhere else in the cosmos, we cannot help but recognize the precious value of our own Earth and the sanctity of the human lives that it supports.

KIGUCHI: In all of human history, not one scientist has ever fully understood what life is or grasped the essential nature of its origins. Since this is the greatest riddle or mystery in the whole of natural science, I believe that scientists in a variety of fields will be challenging this topic from now on.

SHIMURA: Various organic molecules are being discovered, one after another, in interstellar space, by means of the radio-telescope and other highly sophisticated equipment.

KIGUCHI: The first discovery of a hydroxyl group (OH) drew considerable attention, because this molecule requires only one additional hydrogen atom to form water. There was also a great deal of excitement in 1968 when a group of scientists working at the University of California at Berkeley detected the electromagnetic wave pattern of ammonia in outer space. The ammonia molecule has important associations with the life processes.

Proposals for the Twenty-first Century

SHIMURA: Mr. Ikeda, I believe this discovery was made six years before your formal visit to the Berkeley campus where you talked with the chancellor, Dr. Albert H. Bowker. And I believe you also gave a lecture.

IKEDA: Actually the lecture was at the University of California at Los Angeles. I was invited to lecture there by Chancellor Charles E. Young.

KIGUCHI: Under the theme of 'Toward the Twenty-first Century', you made several proposals for the next century. We studied it with great interest, and recently I re-read it. It seems to me that, although ten years have now passed, many of your proposals are becoming increasingly recognized as valid issues.

SHIMURA: A decade at the present time may be comparable, in terms of global communication and the rate of scientific advancement, to several centuries in an earlier era. Nowadays many proposals are outdated by the time ten years have elapsed.

IKEDA: Well, Buddhism has neither beginning nor end (laughter). In that particular lecture, I talked about the Buddhist view of life.

SHIMURA: I hear that the vice-chancellor, Dr. Norman Miller, on hearing your lecture, said that he felt he had glimpsed the essence of Buddhism for the first time. In particular, he said that he gained a deeper understanding of the problem of 'life and death', which modern science is now beginning to confront, and of the Buddhist concept of non-substantiality (*kū*).

Coming back to the subject of the discovery of organic molecules in interstellar space, many nations are presently competing in the effort to discover these basic 'building blocks' of life in space. Thanks to a communications network developed by Bell Laboratories in the United States, such efforts are advancing rapidly.

KIGUCHI: That's true. When ethyl alcohol was discovered in outer space, someone joked that during space travel, one could gather the molecules and enjoy a drink (laughter). Actually about fifty-two molecular compounds have been discovered so far.

SHIMURA: I've heard that recently carbon monoxide has been observed repeatedly.

KIGUCHI: That's right. Carbon monoxide is the organic molecule seen most often. Such observations will help us to clarify the structure of the entire galaxy. Moreover, observations of macromolecular clouds and the behaviour of similar interstellar atomic groupings will contribute to our understanding of the birth of stars. We've also been able to understand in greater detail the distribution of cosmic dust using infra-red telescopes and man-made satellites, such as the one fitted with X-ray observation equipment launched in Japan recently (the *Tenma*, launched from Kagoshima Prefecture on 20 February 1983).[5]

IKEDA: Even though a certain amount of organic materials may have been discovered in space, there is still no sign of life. The unlikelihood of the existence of living beings is so great that even Dr. Sagan confirms that the life span of the universe, as presently estimated, is too short for molecules to have collided a sufficient number of times to produce even so simple a living organism as a virus.

The Unlikely Odds of Protein Synthesis

SHIMURA: Elements such as silicon and germanium are used in the construction of semiconductors, which are widely used in modern appliances. Even if such elements were to collide repeatedly in space, they could not form a transistor even in ten billion or a trillion years. Some scholars cite this example to illustrate the immense odds against life's emergence.

KIGUCHI: Not being an expert in biology, I have no detailed knowledge about the subject, but it is recognized that protein is essential to the structure of living things.

IKEDA: In order for a protein molecule to function, a chain of amino acids must be formed in a certain order.

KIGUCHI: That's right. Simply stated, a single protein molecule consists of a main chain of about two hundred amino acids and twenty different kinds of side chain, a particular type of side chain appearing more than once. The probability of a given protein forming is about one in 10^{200}.

SHIMURA: Ten to the two hundredth power? That's one

followed by two hundred zeroes. If you wrote that figure on paper, you would probably have two or three lines of just zeroes. One hundred million is merely 10^8 and one trillion is just 10^{12}. By comparison, we can gain some sense of the enormous size of the figure you have just mentioned.

IKEDA: In short, the formation of even a single protein, one of life's building blocks, is extremely improbable and virtually impossible to imagine.

KIGUCHI: Mr. Ikeda, you have already mentioned Dr. Sagan's view on this matter, but to elaborate a bit, even if one had tried to synthesize protein once per second in a space of one cubic millimetre, since the time of the Big Bang, he could still not produce a protein of any significance.

IKEDA: And protein is merely one of the factors essential to the emergence of life. The vast numbers and calculations you mentioned are not so much significant in and of themselves as they are as an indication of the extreme rarity of life emerging. And the probability of all necessary factors being present for intelligent life such as human beings to appear is far beyond the scope of scientific speculation. One may argue to no end about whether our presence here is accidental or inevitable, but ultimately we can only call it mysterious.

The Mysterious Nature of the Movements of the Universe

KIGUCHI: What you say is quite true. This is clear in the light of my own field of research, nuclear astrophysics. Perhaps one of the easiest ways to illustrate this is in terms of the distance between the Earth and the Sun. If the Earth were thirty million kilometres either closer to or farther away from the Sun, no life forms, including human beings, could exist here. We are now at the optimum distance from the Sun, one at which water – one of the most essential requirements for the sustenance of life – will neither freeze nor boil in the average atmospheric temperature.

IKEDA: The distance between the Earth and the Sun is about 150 million kilometres. This distance – neither closer nor farther – provides the optimum conditions to give rise to and sustain life. This is indeed a mystic (*myō*) phenomenon (*ho*). *Myō*, among its many meanings, indicates 'the truth'. *Myō* means that which is beyond the scope of the human intellect to comprehend.

If the position of the Earth relative to the Sun were to significantly change, then the Sun's light and heat would cease to support living beings.

KIGUCHI: All living things on Earth depend on the solar energy that they constantly receive from their parent, the Sun.[6] Were the amount of this energy to vary even slightly, it could either an ice age or a great drought. Were it to vary greatly, the entire planet as we know it would cease to exist.

IKEDA: Moreover, if the speed of the Earth's rotation or its revolution around the Sun were to slow down or quicken even a hundred billionth of a second, it could produce a catastrophic earthquake or perhaps even bring about the destruction of the Earth. Thus the first prayer of our morning *gongyō*, performed while facing the eastern sky, has the significance of strengthening the beneficent power and influence of the cosmos which support terrestrial life. Through this prayer we enhance the workings of the universal forces within all life. It includes the phrase from the *Lotus Sutra*, '... day and night, for the sake of the Law, the gods, or the guardians of Buddhism, will ever protect the believer.' We offer this prayer to enable the movements of the cosmos to display their beneficial powers in a precise and continuous fashion.

SHIMURA: Even science cannot explain how various cosmic principles achieve their balance and work together to maintain optimum conditions for the Earth. All science can do is to calibrate their arrived-at interrelation.

KIGUCHI: That's true. There are a great many phenomena which can only be described by the statement that 'the nature of our universe is mysterious,' a remark by Dr. Nigel Calder which is consistent with the understanding of Buddhism.

IKEDA: Were the present gravitational force of our Earth to lessen, people would fly off the surface of the planet. The distance between the Earth and the Sun would alter as well, and the Earth's atmosphere would escape into outer space.

KIGUCHI: Thanks to the law of the conservation of momentum, coins don't go flying out of our pockets to the Moon of their own accord.

SHIMURA: Poverty constitutes a violation of this law (laughter).

IKEDA: That's a matter of fortune accumulated in the depths of one's life, not accessible to the laws of physics (laughter). To give another example, if the principle of the conservation of charge were to fail, the atoms in a piece of paper would all scatter.

KIGUCHI: All the objects we see in our daily lives are held together by the power of electric charges. Were this power suddenly to appear and then disappear, it would be difficult to imagine the chaos.

SHIMURA: A hundred-dollar bill would crumble into fragments before you could use it (laughter).
    So far both of you have discussed the mysterious principles at work in cosmic space, each from your respective standpoints. Seriously, one presently senses a rise in scientific curiosity in the basic trend of our society.

KIGUCHI: In the scientific sense as well, nothing is more mysterious than life itself.

IKEDA: By the way, Dr. Kiguchi, at the beginning of our dialogue you mentioned in passing that there are an estimated ten million planets such as the Earth in our Galaxy?

KIGUCHI: That's correct. A clear evaluation to that effect has been formulated by Dr. Frank D. Drake of the United States.

NOTES: CHAPTER III

1 Mr. Ikeda was named Poet Laureate of Japan by the World Academy of Arts and Culture at the Fifth World Congress of Poets in 1981. The name 'poet laureate' is derived from the ancient Greek tradition of an honoured poet being crowned with a laurel wreath.
2 The tale of Urashima Tarō has similarities with the story of Rip Van Winkle. Although accounts differ considerably, in the standard version, Urashima rescues a turtle being mistreated by children. The turtle turns into a beautiful young woman who escorts him to the sea-god's palace, where he spends three happy years with her. When he finally returns to his village, centuries have in fact passed since he left and he is unable to recognize anyone. Perplexed, he opens a casket that the young woman gave him but has forbidden him to open. White smoke escapes from it, and in an instant he is transformed into a withered old man.
3 *Man'yōshū*: the earliest extant collection of Japanese poetry. 'Leaves', from the English title 'Collection of Ten Thousand Leaves', is the translation of *yō*, which, here, indicates poems.
4 A legendary mountain in China where an elixir for immortality is said to be found.
5 Since most of the radiation from the ultra-violet end of the electro-magnetic spectrum (which includes X-rays) is filtered out by atmospheric layers that surround the Earth (direct exposure to which would cause blindness, and excessive exposure to what little amount filters through the protective ozone layers can result in skin cancer), the use of X-ray telescopes from the Earth's surface is largely ineffectual.
6 Earth and all substantial celestial bodies retain and re-radiate energy received from sources such as the Sun in the form of electro-magnetic energy at the infra-red end of the electro-magnetic spectrum (which is sensibly warm), and this is why the Earth's temperature remains relatively constant even during the 'night' phase of its rotation. The gaseous atmosphere also facilitates the retention of this energy, functioning as insulation, reflecting the warmth back to the Earth's surface. The gaseous compound that is mainly responsible for this insulating effect is carbon-dioxide, which in high concentration have a white or opaque appearance. A sharp increase in this compound, because of car exhaust and other types of industrial pollution, is thought to be responsible for the warming of the Earth's atmosphere, and the irregular weather patterns (cf. the greenhouse effect). It is worth noting that the high concentration of carbon compounds in the atmosphere of Venus and Mercury is, notwithstanding their relatively close proximity to the Sun, chiefly responsible for the extremely high surface temperature of these planets.

IV

Does E.T. Really Exist in the Universe?

Are human beings a solitary existence in the great universe? The latest findings in modern science concerning the existence of 'intelligent life beyond the Earth' confirms the universality of the Buddhist view of the universe.

Another Solar System?

SHIMURA: Last time, we discussed the mystery of the universe in terms of 'life space'. In this session, I hope we can probe a little further into the mystery and wonder of the cosmos. I understand that you once met with Dr. Carl Sagan.

IKEDA: Yes, I've had that privilege.

SHIMURA: Where was that?

IKEDA: At the Sōka Gakkai's International House of Friendship in Tokyo. We talked about various topics for about two hours. It was most informative.

SHIMURA: Dr Kiguchi, I heard that you attended an astronomers' conference.

KIGUCHI: It took place for three days near Tokyo Astronomical Observatory. The topic of the meeting was that a gaseous nebula had been observed within the Orion Nebula (fifteen hundred light years from the Earth) which seems as though it may be capable of generating a solar system.

SHIMURA: It was the first such observation anywhere in the world, wasn't it?

KIGUCHI: That is correct. Astronomers have for some time been puzzled by the two-directional effusion of gases from the vicinity of an unusual infra-red body. After closer observation, it was found that these gases were expanding in a perpendicular discoid fashion.

IKEDA: And that may eventually develop into a solar system similar to ours?

KIGUCHI: Exactly.

IKEDA: Then it is conceivable that a planet similar to the Earth may take form.

SHIMURA: This discovery may then provide a clue to the mystery of how our solar system was formed. Factors essential for the emergence of life in space have been discovered, but as it stands, it is not possible to find living being other than on the Earth. Among astronomers, the idea has taken root that it is easier to search for civilizations in space than to look for living beings themselves; it seems that various research projects and investigations are being conducted along this line.

KIGUCHI: Yes, I believe that was mentioned in an earlier discussion of the project, 'Search for Extraterrestrial Intelligence', or SETI for short.

IKEDA: It would be extremely interesting to know what conclusions present-day science has reached regarding the possibility of the existence of extraterrestrial intelligent life.

The Scale of the Universe as Expounded in Buddhism

SHIMURA: First of all, I would like Mr. Ikeda to explain how Buddhism views this matter.

IKEDA: Well, this subject is very important from the Buddhist perspective on life and the universe.

As we already know, the Earth is not the only object with which Buddhism is concerned. Rather, it takes as its object the entirety of the boundless cosmos, and the ultimate Law of life which pervades it. Mahayana Buddhism, in particular, clarifies the true nature of all phenomena, even including that of a single pebble or article of dust existing anywhere in the great universe, in terms of *ichinen sanzen* or the Mystic Law.

To give one example of how Buddhism expounds the scale of the universe, the *Ninnō Sutra* (*Sutra of the Benevolent King*) states, 'In the realm over which I now reign, there are ten billion Sumeru worlds and ten billion suns and moons, and each Sumeru is surrounded by four continents.' I feel that the expression, 'ten billion suns and moons', deserves close attention, because the idea of infinite expansion is implied by it.

Moreover, in the *Mirai seishukugō senbutsumyō Sutra* (*Sutra of the Names of One Thousand Buddhas of the Future Constellation Kalpa*) and other sutras, we find indications of cosmic time and space.

KIGUCHI: May we interpret the 'four continents' as the realm inhabited by intelligent life similar to human beings?

IKEDA: I think so. I understand that astronomers have estimated there may be as many as ten billion planets with conditions able to support civilizations like the Earth's in the Milky Way alone.

KIGUCHI: Yes; if you bear in mind the famous equation which the American astronomer Frank Drake of Cornell formulated and Carl Sagan publicized in his book *Cosmos*, the number is indeed prodigious.

The Universal Propagation of the Mystic Law

IKEDA: Nichikō Shōnin (1867-1957), the fifty-ninth high priest of Nichiren Shōshū, was an unrivalled scholar of Buddhism. I had the honour of meeting him several times, while accompanying my teacher, Jōsei Toda.

On one occasion, when we were discussing the sequence of Buddhist propagation in terms of the 'five periods and eight teachings', the high priest made the following remark: 'On this Earth, the stream of Buddhism began with the Tripitaka teaching, proceeded to the connecting, specific and perfect teaching, and ultimately, *Nam-myōhō-renge-kyō* was revealed. It seems to me that, after this is established on Earth, a similar process may occur on other planets, I remember this statement of his vividly and it still holds much interest for me.

SHIMURA: In other words, first Shakyamuni appeared in India and expounded *zōkyō* (the Tripitaka teaching), *tsūgyō* (the connecting teaching), *bekkyō* (the specific teaching), and *engyō* (the perfect teaching). Then in the Latter Day of the Law, *Nam-myōhō-renge-kyō*, the fundamental seed of Buddhahood, was revealed. And Nichikō Shōnin conjectured that this process is taking place on other planets as well.

IKEDA: I recall Mr. Toda making a similar remark in his address at a general meeting of the Sōka Gakkai youth division in the 1950s. At that time, its significance was far beyond our imagination, and we could only interpret it as a humorous remark on his part.

His lecture ran along these lines: 'On this planet, the Buddhism of Nichiren Daishōnin has been firmly established. I myself have now opened a path toward the fulfilment of *kōsen-rufu*, or the "widespread dissemination of the Law". I will now leave the rest to you, as I must go on to some other world to engage again in the task of propagating the Mystic Law.'

KIGUCHI: Mr. Ikeda, you have outlined the Buddhist view on this subject in an easily understandable manner. It is only in the twentieth century that science has begun to arrive at the high level of understanding reached long ago by the Buddhist teachings with respect to the notion of galactic systems in which countless solar systems and life-supporting planets are thought to exist.

IKEDA: Come to think of it, it was only three hundred and fifty years ago that Galileo asserted that the Earth is no more than one of several planets orbiting the Sun.

KIGUCHI: Galileo observed, with a telescope he had adapted to his needs, that the Milky Way is a cluster of innumerable stars. It appears, however, that he had not yet realized that there exist myriads of solar systems similar to ours.

Earth: A Spaceship with 4.5 Billion Passengers

SHIMURA: It appears that astronomy progressed by trial and error for a long period of time before arriving at the concept of the galactic system model, which is only now being clarified.

KIGUCHI: There were, simply speaking, two major turning points along the course of astronomy's advancement. One occurred at the turn of the nineteenth century, when the British astronomer Sir Frederick W. Herschel (1738-1822), together with his sister and son, also astronomers, turned his telescope to the solar system and provided some clues to the structure of the Galaxy. The second was the achievement of the American astronomer, Edwin Hubble (1889-1953).

SHIMURA: I believe he discovered 'the Hubble constant', which became the basis for calculating the theory of an expanding universe.

KIGUCHI: Hubble observed nebulae beyond our galactic system, and ascertained the size of the Galaxy with respect to the whole of the universe. This was only slightly more than fifty years ago, but astronomy has taken a giant step forward since that time, thanks to his efforts.

SHIMURA: As our understanding of the galactic universe has progressed, its characteristics, laws and physical properties have become clear. And the question of whether man is the sole intelligent inhabitant of the universe has developed into a major theme.

IKEDA: Whether life in outer space exists or not is a matter of tremendous concern for human beings. Legends like those of the

Shining Princess[1] or the Rabbit on the Moon[2] represent a world of fantasy that has been passed down through countless generations in human history. Such legends seem to illustrate that people have always had a deep longing for space, regardless of the age.

KIGUCHI: In that sense, astronomy represents the human being's never-ending spirit of inquiry.

IKEDA: The notion that the Earth is like a spaceship, orbiting the Sun with 4.5 billion passengers on board, has won acceptance only by virtue of our expanding notions about the nature of the universe.

SHIMURA: This is an extremely important perspective for humanity to acknowledge. Unless we can continue to strengthen our awareness of the common destiny of all who live on this planet and of the irreplaceability of the Earth itself, we might witness the end of the human race. Moreover, I think it will become essential for intellectuals and world leaders to investigate Buddhist thought. Because it embodies wisdom which enables us to understand concepts which science has just begun to elucidate, Buddhism can serve as the basis for the kind of sound decision-making and planning based on foresight that is needed to cope intelligently with the current problems of unprecedented immensity – problems that demand speedy resolution to avert calamity.

Ten Million Civilized Societies May Exist in the Galaxy

KIGUCHI: The Buddhist concept *sanzen daisen sekai*, or a major world system, is quite impressive.

IKEDA: The *Juyrō* (sixteenth) chapter of the *Lotus Sutra* refers to 'five hundred, thousand, ten thousand, hundred thousand, *nayuta, asōgi* major world systems', a view of cosmic vastness beyond our imagination. To elaborate, the numerical unit *nayuta* has several calculations. According to one, it equals one hundred billion in our present numerical system. An *asōgi* is often defined as 10^{51}, but for all practical purposes it indicates a large, virtually incalculable number.

SHIMURA: Today astronomers, space biologists and others in related fields are forming hypotheses and conducting various investigations in the hopes of determining the possibility of extraterrestrial intelligence, based on a variety of data.

KIGUCHI: American scholars, who in 1959 proposed a project to attempt communication with intelligent life in space, said that 'the probability of success is difficult to estimate, but if we never search, the chance of success is zero.'

IKEDA: That indeed is the attitude of a pioneer.

KIGUCHI: Project OZMA,[3] a search for signs of extraterrestrial life launched in 1960 by the National Radio Astronomy Observatory (NRAO) in the United States, was the first such attempt to be carried out anywhere in the world.

SHIMURA: Have there been other attempts since then?

KIGUCHI: I believe there have been about six or seven major projects, including those undertaken by the Soviet Union. Project Cyclops,[4] supported by the National Aeronautics and Space Administration (NASA), is said to be a full-scale operation with a budget of about twenty billion dollars.

IKEDA: Do these search projects share a common conception of how such beings might be located?

KIGUCHI: Yes, it seems so. In simple terms, they search for fixed stars like the Sun, and then consider the possible existence of planets in orbit around those stars with conditions similar to those of the Earth.

SHIMURA: How many such planets are estimated by astronomers to be within our Galaxy?

KIGUCHI: Opinions differ, but a representative view would be that of Dr. John Kraus, head of the Wesleyan Radio Observatory in Ohio, USA.

IKEDA: Would you explain his views to us in specific terms?

KIGUCHI: He says in essence that the Sun is but one of a hundred billion stars which constitute the galactic system, and the galactic nebula is but one of the hundred billion nebulae which exist in the universe. Even postulating only a million-to-one chance that stars with planets like the Earth exist, there would be some one hundred thousand such planets in our Galaxy. However, since astronomers conjecture the existence of more stars than this, it is quite conceivable that somewhere in the Galaxy life does exist.

IKEDA: In our galactic system alone, there exist several billion fixed stars like the Sun, with a great number of planets orbiting them. The number of such planets in the entire universe is beyond reckoning. The great majority of them can be assumed to support no life whatsoever. However, I have heard that those planets, which have a relationship to their respective stars similar to that of the Earth to the Sun, may possibly have the conditions necessary for life to emerge. What is your opinion of this, Dr. Kiguchi?

KIGUCHI: Given our current understanding, it is an entirely reasonable view.

SHIMURA: I imagine the number of possible life-supporting planets would vary in accordance with the equation one uses in calculation. Dr. Kiguchi, what is the equation of Dr. Drake which you mentioned earlier?

KIGUCHI: I will abbreviate the calculations, but the equation is as follows: $N = N_* f_p n_e f_l f_i f_c f_L$. N indicates the number of technically advanced civilizations in the Galaxy. N_* represents the number of fixed stars born within the Galaxy in a given unit of time; f_p indicates the percentage of fixed stars having a planetary system; n_e is the number of planets among those planetary systems having conditions suitable for life; f_l is a fraction of those planets on which life actually arises; f_i is the percentage of those planets on which life has evolved into intelligent forms; f_c is a fraction of planets on which intelligent beings have developed a technical civilization to the point where they can aim at interstellar communication with, for example, the Earth; L represents the life span of those technically-advanced civilizations.

IKEDA: Then one must supply highly detailed probabilities for each factor.

KIGUCHI: That's correct. Dr. Drake's equation attempts to calculate this probability. At present we can make comparatively accurate conjectures for certain factors, but the majority are still unknown.

IKEDA: Has any scientist formed a rough estimate based on that equation?

KIGUCHI: Yes, a number of scientists have made estimates, based on the knowledge of their respective fields of specialization. In Japan, the conjectures of a biochemist, Dr. Tairō Ōshima, have been published in his book *Seimei no tanjō* (*The Birth of Life*), and are generally well-known. According to his estimates, no less than ten million advanced technological civilizations could exist in our galactic system. However, from a biological standpoint, the survival of a given species is estimated at ten million years. If we assume that the human race will endure for this length of time, the factor N, the number of highly advanced technological civilizations in the Galaxy existing contemporaneously with the human race, would be estimated at about one hundred thousand.

Discovering Fixed Stars Like the Sun

SHIMURA: Incidentally, about fourteen years ago, the

American research institute, Rand Corporation, announced that within thirty years, humanity might be able to carry out two-way communication with extraterrestrial beings. Concretely, how are such investigations conducted?

KIGUCHI: Researchers listen for particular radio waves from other stars. The participants are paying special attention to those stars which could possibly have planets with civilizations far more advanced than that of the Earth.

IKEDA: Assuming the age of the Earth to be 4.5 billion years, if we take a planet which was born one per cent earlier, it could have a civilization forty-five million years ahead of ours.

KIGUCHI: It is, therefore, conceivable that such a planet may have a civilization advanced enough to conduct interstellar radio communication, possibly in the interest of locating other planets with intelligent life.

SHIMURA: About a dozen years ago, there was talk that someone had picked up Morse code-like signals from outer space. It raised quite a furore.

KIGUCHI: A student majoring in astronomy at the University of Cambridge discovered it by chance. It seems that some astronomers at first thought that intelligent life on another world might be sending a message to the Earth. They dubbed the planet LGM, for 'little green men', but it was proved almost immediately to be a pulsar, a small, dense star with a powerful magnetic field that emits radio waves as it rotates. Scientists are grappling with this problem of extraterrestrial intelligence in earnest, based on precise calculations and research, setting aside all biases and preconceptions. Dr. Anthony Hewish, the above-mentioned student's professor, won the Nobel Prize for his contribution to the study of pulsars.

SHIMURA: You mentioned earlier that the search for extraterrestrial intelligence begins with the search for fixed stars like the Sun.

KIGUCHI: That's right. When the light from these stars is passed through a prism, we can judge their temperature and approximate age from the emission and absorption lines of the resulting spectrum.

IKEDA: How are fixed stars identified?

KIGUCHI: We use a classification system called *spectral type*. When stars are classified according to this system, large, hot stars are called blue, and smaller stars with a lower temperature are called red.

IKEDA: Then what colour is the Sun?

KIGUCHI: The scale is divided into ten grades – from Class O stars to Class S – and our Sun is right in the middle, categorized as yellow.[5] It is conceivable that other yellow stars like the Sun may have planets with intelligent life.

SHIMURA: How far away is the nearest star in the same class as the Sun?

KIGUCHI: Approximately four to ten light-years.

IKEDA: Then, even if the Earth were to pick up radio waves from a star in that solar system, they would have been sent ten years ago.

KIGUCHI: Yes, that's a problem. Another is that we must keep our radio-telescopes trained in exactly the right direction. Moreover, among countless possible wavelengths, we must tune in on precisely the ones that can be used for communication. To compound the difficulty, there are times when we can see the 'antenna' – the star itself – and times when we cannot.

IKEDA: That is because of the rotation and revolution of both the star and the Earth, is that right?

KIGUCHI: That's correct. The observations must be conducted without interruption for even a single moment.

SHIMURA: According to a recent news report (8 March 1983), the American scientists have developed a radio-telescope capable of analyzing in the space of one minute whether or not a particular signal is a radio wave transmitted from space.

KIGUCHI: In the United States, through the use of computers, the ability to detect whether or not such signals carry intelligible meaning has greatly advanced.

Even with this telescope, however, the probability of finding intelligent life is extremely small, because likely fixed stars are difficult to locate. Nevertheless, many astronomers find it appropriate to think that planets where civilisation thrives do indeed exist.

Our Bodies Come from the Stars

IKEDA: About eleven years ago, I believe, a rocket aimed outside the solar system was launched. Is it still flying smoothly?

KIGUCHI: Yes, indeed. You refer to the unmanned space probe, *Pioneer 10*. It was recently confirmed that it has finally crossed the orbit of Neptune and is continuing on its flight into outer space.

SHIMURA: For approximately how many more years will NASA be able to receive signals from this rocket?

KIGUCHI: The atomic-powered battery loaded on board is expected to last another twenty-four to twenty-five years.

IKEDA: Then it is possible that *Pioneer 10* will transmit new information about the universe to us.

KIGUCHI: At any rate, this is the first man-made device to venture out of the solar system, so we are eagerly awaiting its findings.

SHIMURA: I understand that the rocket carries a message

written in 'space language' in the event that extraterrestrials may discover it.

KIGUCHI: That's right. It's a message devised by Dr. Sagan and others, communicating various data about the Earth.

IKEDA: I hear we have also transmitted various signals to outer space. Does there seem to be any likelihood of a response?

KIGUCHI: Considerable difficulty is involved. The stars are so far away that it would take several dozen or perhaps even more than a hundred years for an answer to arrive. However, scientists are working in earnest to devise combinations of symbols, pictures and numbers that intelligent extraterrestrial beings might be able to understand.

SHIMURA: Some scholars feel that the Earth is not particularly unique, and that it is therefore reasonable to assume that other planetary environments would exhibit the same general characteristics.

KIGUCHI: Many scientists believe that the laws of physics and chemistry are universal principles which should apply anywhere in the cosmos.

IKEDA: All living things on Earth are composed of about twenty basic molecules, so, probabilities aside, it is quite natural to think that extraterrestrial life might also be composed of the same molecular compounds.

KIGUCHI: Astronomers often say that our bodies have come from the stars. In other words, our bodies are composed of substances that were formed in stars. The most essential element for the formation of life is carbon. Much research has been done on this subject, but it appears that such elements as silicon and tin are simply incapable of forming enough diverse chemical compounds to sustain life. Incidentally, carbon is the element most readily produced in the interior of stars. Why this should be so is still a mystery.

What the Earth Lacks: The Purification of the Six Sense Organs

SHIMURA: Incidentally, Dr. Kraus of the United States, whom we mentioned earlier, also seems to be involved in the project of communication with outer space. He has stated that if there are planets older than the Earth which have developed and sustained a high degree of civilisation for several hundred thousand or millions of years, their inhabitants must be exceptionally merciful and tolerant.

IKEDA: If that is the case, then in Buddhist terms, we would say that those inhabitants must each have deep faith and the world must be a place enveloped in compassion, where *kōsen-rufu* has been achieved.

SHIMURA: In his book *Life in the Stars*, the British author Sir Francis E. Younghusband (1863-1942) depicts the denizens of outer space as creatures of a very high order, possessing 'angelic dispositions'.

KIGUCHI: I believe that is an idea shared by many scientists.

IKEDA: Speaking again in Buddhist terms, we would say that such beings have achieved the purification of the six sense organs.

SHIMURA: The six sense organs comprise the five sensory organs – eyes, ears, nose, tongue and body – as well as the mind.

IKEDA: The expression 'purification' of the six sense organs might seem to suggest secluding oneself in the mountains and practising austerities to eradicate desires, but this is not what it means. The true purification of the six sense organs is achieved by invoking the Mystic Law and fusing one's life with it. In short, it may be called the purification of life itself.

SHIMURA: Along with the advance of science and technology, human life itself must also be cultivated.

KIGUCHI: Intelligent extraterrestrials who have long sustained a highly developed civilization – if they exist – may be assumed to have developed into beings of outstanding character. Had they not, they would have brought about by their own hands the destruction of their civilization through egotism and arrogance.

Religion of and for Human Beings

SHIMURA: The *New York Times Magazine* (23 January 1983) carried an exclusive interview with Dr. Stephen W. Hawking, who is said to be a second Einstein. He was quoted as saying, '... the more we examine the universe, we find it is not arbitrary at all but obeys certain well-defined laws that operate in different areas. It seems very reasonable to suppose that there may be some unifying principles, so that all laws are part of some bigger law. So what we are trying to find out is whether there is some bigger law from which all other laws can be derived.' Couldn't we say that this view represents a step forward for scientists challenging the riddle of the universe?

KIGUCHI: I think so. Astronomy at present has yet to shed light on the universe before the Big Bang. Dr. Hawking is currently speculating on what may have produced the Big Bang.

IKEDA: 'Some bigger law from which all other laws can be derived' is certainly suggestive. Buddhism teaches that 'infinite meaning derives from the one Law' (*Muryōgi Sutra*). This 'one Law', ultimately, is *Nam-myōhō-renge-kyō*.

SHIMURA: Dr. Hawking, while suffering a debilitating and incurable disease, holds the Lucas Chair[6] at the University of Cambridge, which Newton once held.

IKEDA: His pursuit of a single truth even while in impaired health is truly worthy of respect.

SHIMURA: I once interviewed Dr. Jacques Monod (1910-1976)

of France, molecular biologist, then head of the Pasteur Institute. This was shortly after he was awarded the Nobel Prize.

IKEDA: He wrote a book-length essay called *Le hasard et la necessité* (1970; *Chance and Necessity*, 1971).

SHIMURA: At the time of our interview, I noticed he had an esoteric Buddhist icon on the wall of his library, so I asked him how he had come to take an interest in Buddhism. He remarked how difficult it is to clarify life itself. He said that in the light of his speciality of molecular biology, even if he were able to assemble all the necessary materials and equipment, the odds against being able to produce even a single one of the proteins needed for life to form are astronomically great.

KIGUCHI: Scientists are often struck with awe and wonder in the face of the mysteries of life and the limitations of science, and feel almost inclined to bow their heads.

IKEDA: Buddhism speaks of *honchi nanshi*, which means the 'original entity which is beyond comprehension'. The 'original entity' of the Buddha is *Nam-myōhō-renge-kyō*. *Nam-myōhō-renge-kyō* is the ultimate Law which manifests and supports every function of life. It is also called the Mystic Law.

SHIMURA: Dr. Monod also mentioned how fortunate the Earth is in terms of its location in the universe, and said, 'Science can explain the composition of things, but it cannot possibly explain why they came into existence, or why they are as they are. Were it to attempt to do so, it would have to abandon its premise of objectivity.' He also said that if one seeks the answers to these questions in Christianity, no matter how hard he may seek, he can only move farther and farther away from the scientific 'premise of objectivity'.

KIGUCHI: His way of thinking would appear to represent a major turning point in the history of science. Dr. Monod insisted that one should not invest science with one's personal value system, but he himself is a pioneer of the modern trend to unite human values, and also religion, with science. I recall that in your interview, carried in some issue of the *Daibyakurenge*[7] (February 1973), he said something to the effect that Buddhism possesses concepts of values and ethics not found in science, and expressed his hope for 'religion of and for human beings'. Upon reading that article, I felt strongly encouraged about the future of science.

An Increasing Trend toward Buddhism

SHIMURA: To change the subject, are there scriptures which describe the method of propagation of true Buddhism?

IKEDA: In the *Shrimala Sutra* (a sutra translated into Chinese in 436 by Gunabhadra), in the second passage of Chapter Ten it states: 'When preaching Buddhism, employ *shōju* to people when it is appropriate and employ *shakubuku* to people when it is appropriate. These two methods should be used according to the knowledge and understanding of Buddhism on the part of the listeners.'

Also, in a scripture called the *Fugen Sutra*, which is the conclusion to the *Lotus Sutra*, it is explained: 'Should you wish to do *shakubuku*, you must do so with the teachings of Mahayana Buddhism.'

Furthermore, in the *Hokke gengi* (*Profound Meaning of the Lotus Sutra*), a lecture in ten fascicles by Chih-i, also known as the Great Teacher T'ien-t'ai, elaborating on the essence of Buddhism, it states: 'The *Lotus Sutra* is the teaching of *shakubuku*, the refutation of the provisional doctrines.' We carry out propagation activities accordingly.

SHIMURA: Many claim, based on the superficial interpretation of the word, that *shakubuku* is a method of extreme intolerance.

IKEDA: I hear that quite often, but the word *shakubuku* was not coined by us. The word was used by Shakyamuni Buddha and Nichiren Daishōnin as a general term which applied to Buddhism as a whole. We use it today in the spirit of carrying out its true meaning.

Shakubuku means to eliminate evil and to cultivate a heart of virtue. Lower forms of religion or faith weaken the life force of people, hinder the development of civilization, and obscure the true meaning of life. The reason for our doing *shakubuku* is that, based on mercy, we want mankind to cultivate the highest level of life, a well-developed personality, and the most highly-evolved perceptions of society, the world and the universe, through faith in the essence of Buddhism. In so doing, we must elevate ourselves from the lower to the higher planes and, by penetrating our own shallowness, become a person of depth. In other words, one becomes an individual of great tolerance and magnanimity.

SHIMURA: It is often said that the Buddhism of Shakyamuni is like the benevolence of a mother, and the Buddhism of Nichiren Daishōnin for the Latter Day of the Law is likened to the strict affection of a father.

IKEDA: In observing the increasingly decadent and corrupt social conditions in today's world, propagation of a religion which is only half-hearted and not truthful will not penetrate the lives of the people. The propagation of Buddhism which espouses the ultimate Law of the universe, carried out with great conviction and backed by the eighty thousand teachings, may seem strict and relentless, but it is my belief that this is the method that is most appropriate in the Latter Day of the Law.

This is an age when people of authority – school teachers, parents, religious leaders and politicians – are no longer respected. I am reminded of the passage in the *Hōon shō* (*Repaying Debts of Gratitude*) which says, 'To discard the shallow and seek the profound requires courage.'

KIGUCHI: I have read somewhere that teachers who are weak in character are usually victims of violence in elementary, junior high, and high schools.

IKEDA: Many problems involving young people arise from the fact that there is no place for them to go and no one to rely on or to guide them any more as they face the future. In the past, they used to be able to depend on teachers, religious leaders and doctors. In recent years, however, the spirit of the young people has changed so much that they no longer trust anyone in a position of authority.

It is only natural, therefore, that human beings who are underdeveloped spiritually, and society as a whole, will allow their impulses to take over and run wild, without the guiding hand of the older generations to offer them solid direction.

SHIMURA: This phenomenon is nothing less than terrifying.

IKEDA: A number of eminent people are debating this issue and groping for ways to solve these problems. This, of course, is important, but the ideal is to seek solutions through concrete actions based on Buddhist principles, on a day-to-day basis, tackling each and every problem, rather than engaging in abstract discussions.

NOTES: CHAPTER IV

1. Also known as *Taketori no okina no monogatari* (*The Tale of the Bamboo Cutter*), an early Heian-period (794-1185) prose work about a supernatural being found in a bamboo stem by a bamboo cutter and brought up as his daughter. Eventually, she reveals to her parents that she is from the Palace of the Moon. She then dons a robe of feathers that obliterates her memories of the world and she departs for the Moon.
2. It was believed that a rabbit pounding rice could be seen in the Moon when it is full.
3. A research project to find higher forms of life carried out in the United States as a part of the space program. The team attempted to analyze the radio waves from outer space and to extract signals from them. However, there were no signals which would indicate the existence of extraterrestrial beings.
4. The SETI project which was announced in 1971. Over one thousand antennae, of 100 metres in diameter, will be controlled by computers, the expense being approximately twenty billion dollars.
5. The classification system most often used currently is the Yerkes system, with the following spectral types: O, B, A, F, G, K, M, R, N and S. Classification by spectral type is essentially a temperature classification, Class O stars being the hottest, with surface temperatures of up to $40,000°K$. The Sun is a typical Class G star of about $6,000°K$. The cool red stars of classes M, R, N and S have temperatures below $4,000°K$.
6. A professorship of mathematics, optics, etc, established at the University

of Cambridge in 1663 through the donation of Henry Lucas. It was first held by Isaac Barrow, under whom Newton studied, and Newton was the second to assume the position at the age of twenty-six.

7 The Sōka Gakkai's monthly Buddhist study magazine.

V

Buddhism, the Universe and Human Existence

A Buddhist dialogue on the dynamics of universal phenomena and human existence.

The Marvellous Women Cosmonauts

IKEDA: Nichiren Daishōnin states, 'Buddhism is reason.' We may interpret this to mean that what brings about a balance which is essential to both the individual and society is Buddhism itself. People may discriminate against others because of differences in race, social systems and ideologies, but human beings themselves are the common denominator in any society.

KIGUCHI: No matter what society or country, the joys and sorrows of human beings everywhere are the same.

IKEDA: Looking back over the many thousands of years of human history, happiness and sorrow, however great or small, alternate in an unending cycle. Social systems as they were a thousand years ago differ greatly from those of the present day, and no doubt, they will change completely in the future. Change is wrought by the passage of time, and the source of change lies in human beings. Therefore, irrespective of any age or country, the deeper the people's anxiety and despair, the greater their longing for peace and happiness. This is human nature, and the principle which grasps the true nature of man and provides the means to realize his desires is 'reason' as expounded by Buddhism.

SHIMURA: In Japanese, the word 'reason' is *dōri*, also meaning 'principle' or 'truth'. Though we often use it on a daily basis, in Buddhism it has a much deeper meaning.

IKEDA: The Buddhist sutras divide 'reason' or 'principle' into four categories. To explain briefly, they are principles of relative relationships among all phenomena, principles of functions based on causality, principles which are demonstrated through inference and inherent principles in all phenomena.

KIGUCHI: When we discuss matters from the standpoint of Buddhism and the universe, which category are we employing?

IKEDA: The category of inherent principles, because Buddhist teachings are expounded as principles relative to the Law, underling all phenomena of the universe. Many words which the Japanese use in daily conversation originally derive from Buddhism. In Buddhism, the word *dōri* embodies profound perceptions concerning the relationship between the human being and the cosmos. In this sense, we can say that Buddhism aligns itself with reason. In the course of its nearly three-thousand-year history, however, there have also been cases in which it became ritualized, its original spirit lost and its form remaining only as a part of tradition. It is, therefore, important to make continuous efforts to restore and revitalize Buddhism, shedding light on its true essence.

Religion, which should be the basis of everything, must necessarily undergo a kind of revolution of its own, while basing itself on its original spirit and tradition. The strong commitment of those who uphold it toward their own inner reformation becomes crucial and indispensable.

SHIMURA: Incidentally, Ms. Sally Ride, the first American woman astronaut, successfully carried out her mission on the space shuttle *Challenger* during its seventh flight. When was it that you met Russia's female cosmonaut, Ms. Valentina V.N. Tereshkova in Moscow?

IKEDA: Around May 1975.

KIGUCHI: I recall reading your impressions concerning that meeting in a book entitled, *Wasureenu hitobito* (*Unforgettable People*), and about how Ms. Tereshkova, of humble origins, had gone from being a girl who worked in a factory to a kind of space heroine, having undergone severe training in order to carry out her mission.

IKEDA: She was totally unaffected. I thoroughly enjoyed speaking with her, and we both regretted not being able to talk further because of lack of time. Actually, our meeting was not included in the itinerary of my trip to Moscow. What remains fresh in my memory, however, are her words, 'Anyone who has ever seen the Earth from outer space looks at the cradle of human life as being truly precious and dear.'

SHIMURA: It is worth mentioning that Ms. Ride was trained to be one of several technicians along with other astronauts, mostly men, in an ambitious space programme. Although she is a scientist, she was trained to be able to pilot the shuttle in case of an emergency.

KIGUCHI: As I watched the shuttle in flight on television with great interest, I realized that space exploration is no longer restricted to men only.

IKEDA: Time passes, changes unfold and further advancements are made. Soon it will be considered commonplace for women to travel in, and explore, space. The smaller the Earth becomes, the closer the universe will also become. For this reason, we need a philosophy which encompasses both the Earth and the heavens, clarifying the mutual relationships of all things. And the highest priority should be given to human beings, which is possible through the teachings of Buddhism.

KIGUCHI: Otherwise, people will become like robots. The advances of science should not result in the dehumanization of human beings. Along with the flow of the times, systems of thought must become more relevant to humanity; if not, any scientific development will be meaningless.

IKEDA: Although Ms. Tereshkova's serene manner belies her strong character, she has succeeded in her own way in realizing a kind of human revolution. Ms. Ride as well is independent and considers herself to be an American woman who is always active and on the go. These two women exhibit the spiritual tenacity which enables a person to overcome difficulties with grace and motivate others, and strong commitment and aspiration which reveal the magnificence of the human spirit.

Touching on the Heartstrings of the Universe

SHIMURA: A reader wrote to our magazine recently, saying, 'This summer, I felt a different emotion as I looked up at the starry sky.'

KIGUCHI: While having to make our way through the complexities of human relationships in our society, it is indeed with a sense of relief that we look up at the heavens.

IKEDA: Recently, news concerning space has become more commonplace. For instance, the discovery of a new comet and the occurrence of a solar eclipse have been much reported in the press. I would imagine, Dr. Kiguchi, that you are very busy these days.

KIGUCHI: In the study and research of space, observations, investigations and calculations require a considerable length of time and patience.

SHIMURA: Coverage of findings or events by the mass media sometimes results in chaos. For example, in June 1983, about one thousand Japanese people visited Java, an Indonesian island, in tour groups just to have a better view of the solar eclipse, and some even made reservations over a year in advance.

IKEDA: It's a good sign that people are taking more interest in what goes on in space and in the latest scientific developments, but such interest should not be mere idle pursuit, resulting in little in the way of value creation. We should cultivate our curiosity in the interest of becoming more sensitive to the infinite workings and driving rhythm of the universe.

SHIMURA: Recently a kind of space 'fever' has been taking over, particularly on television. There are as many as twelve space programmes for children. They are mostly weekly cartoon serials, so the children monopolize the TV in the evening. Dr. Kiguchi, as an astronomer, how do you feel when you look up at the heavens?

KIGUCHI: Every day, because of my work, I am always preoccupied with the relationship between the planet Earth and the universe. But there are moments when I am suddenly overwhelmed by a sense of awe. We may very well be able to depict the Earth with numerical figures and graphs, but it is, nevertheless, very difficult to grasp its substance and tangibility.

IKEDA: One would think that the Earth has been thoroughly studied; there are no new continents to be found, for instance. But it is paradoxical to me that the planet is riddled with warfare and conflicts, particularly at a time when global unity among all peoples is most needed. Since no country can survive independent of the rest of the world, in these times of global crises, a life-philosophy capable of uniting the world should be sought after with an even stronger sense of urgency.

SHIMURA: The times demand the continuous exchange of friendship on a one-to-one basis among all nations.

IKEDA: After all, it is the human being's wisdom which decides the destiny of any country or society. Just as the field of astronomy is making steady progress toward touching on the 'heartstrings' of the universe, we as human beings, who are like microcosms, must strive to reach out to each other with utmost sincerity, deepening mutual trust and understanding.

KIGUCHI: I remember these words which you wrote: 'Among all human beings, there are heartstrings which vibrate and a chord which strikes sympathy and resounds in the beautiful harmony of life-to-life communication.'

SHIMURA: Those words are indeed beautifully written. Such poems are as numerous as the stars in the heavens, but few of them are so penetrating, appealing and based upon one's action.

KIGUCHI: The study of astronomy would seem to be far removed from human sentiment.

IKEDA: But every discipline has its own *raison d'être*. Buddhism expounds the concept of 'the perfect manifestation of the entity of all universal phenomena, especially that of the individual human being's life', meaning that all things in the universe, illuminated by the Mystic Law, reveal their own potential and inherent value.

KIGUCHI: There is an anecdote about a Japanese astronomer who was so wrapped up in his work that he did not even know about the outbreak of the Russo-Japanese War in 1904.

IKEDA: Even Newton is reported to have been so absorbed in his thoughts that he once boiled his pocket watch instead of an egg.

SHIMURA: There have been many scientists who were dominated by their passion for discovery or a new invention.

IKEDA: There are people who are obsessed, but aside from these exceptions, the purity and earnestness with which scientists pursue the truth are most worthy of respect. We must never forget how much their devotion has contributed to human happiness. Ignorance of worldly affairs on the part of a great scientist may in some sense be regarded as a virtue.

KIGUCHI: What you say has relevance because there is a tendency for astronomers, for example, to develop a consciousness of only the celestial bodies, concentrating all their mental energies in that direction. One eventually judges the world from the standpoint of his particular, narrow viewpoint.

IKEDA: It seems that the more earnest a scientist is, the easier it is for him to become deceived and exploited.

KIGUCHI: That's correct, and as a matter of fact, Albert Einstein and Bertrand Russell initiated their own peace movements after having reflected on their unguarded and imprudent attitudes as scientists toward the use of technology.

The Ancients Were Great Astronomers

SHIMURA: Most of the ancient legends and myths were connected with ideas about the universe.

IKEDA: It's natural, I think, that the people of those times, who lived more in harmony with nature, should have felt great awe for it, always striving to cultivate a kind of communion between human beings and the universe. Logic and theory held little importance for them. The ancient religions which gave meaning to existence evolved around intuitive wisdom. Every morning they greeted the brilliant morning sun and every evening bade the setting sun, by this time a glowing red ball, farewell. They communicated with the glittering stars set against the blackness of the night sky, pondering the wonders of the universe. Surely they must have arrived at profound realizations about the underlying truth of the great cosmos. Even without any scientific knowledge, the people could experience the benevolence of the Sun in their daily lives. It is often the case that once a person becomes knowledgeable, he tends to become arrogant and then his sense of judgment becomes clouded.

KIGUCHI: In many cultures, including Eskimos and the Japanese, sun worship was an integral part of the people's lives. It is easy to understand why.

SHIMURA: We can say, therefore, that theoretical knowledge does not necessarily enable one to understand the truth of things.

IKEDA: In a sense, one can intuitively perceive the truth based on his own subjective experience.

KIGUCHI: This idea can apply to understanding a great man. Even without having read his books or made careful studies of his ideas, observing his behaviour can enable us to deepen our trust and respect for him.

SHIMURA: However, there are instances where, by coming to understand someone in this way, we also become disillusioned.

KIGUCHI: The ancient peoples observed the Sun closely and as a result knew when the seasons changed. They could observe the movement of the constellations, and they also noticed that solar and lunar eclipses occurred in cycles that were discernible. Consequently, based on their understanding, they were able to govern their lives in rhythm with these phenomena.

SHIMURA: That was the beginning of civilization.

KIGUCHI: It is thought that the study of astronomy began around that time.

SHIMURA: One could say, then, that in ancient times, everyone was a kind of astronomer.

IKEDA: Whether they were hunters, nomads, farmers or fishermen, they had an acute perception of the movement of the stars. Their senses were finely tuned. They may well have been great scientists and astronomers. They knew when to sow the seeds, when to hunt and graze cattle and what the best times were to go fishing. It was crucial for their survival to be able to draw forth their innate wisdom.

KIGUCHI: One eternal truth is that the human being cannot live separate and apart from the environment, and neither human beings nor the Earth itself can exist apart from the universe.

IKEDA: It is my understanding that the English word 'consider' derives from 'star', meaning 'along with the stars'. The word is said to result from the keen understanding of the relationship between man and the stars. It is no exaggeration to say that such a coexistence has been a reality since the beginning.

KIGUCHI: It was a fact that if one miscalculated the appropriate time to plant crops, there would be no harvest and starvation would be the result.

IKEDA: It was precisely for this reason that astronomical observation became extremely important. On the other hand, those in positions of power would monopolize this responsibility, sometimes acting as priests. A ruler, for example, would gather his people in a temple for the purpose of announcing future astronomical occurrences, such as eclipses. In this way, he took advantage of them in the interest of maintaining his own power.

KIGUCHI: Well known is the Temple of Karnak, near the Nile River, in Egypt. It was the centre of religion and government at the height of the ancient empire. Stonehenge, which is located on the plains of Salisbury in England, is the site of the ancient stone monument of the Neolithic era, made of giant stones. Being much talked about recently, it is thought to have been an observatory for watching the stars.

SHIMURA: About twenty years ago, Dr. Gerald S. Hawkins, an astronomer, published an article in the science journal, *Nature*, entitled 'Stonehenge Decoded'. It apparently caused a great stir.

KIGUCHI: It resulted in a new branch of study, 'archaeo-astronomy'.

SHIMURA: When the ancient Japanese tomb Takamatsuzuka was discovered in March 1972, a map of the stars was found drawn on its walls. Because of the existence of that map, it has been surmised that astronomical observations in ancient Japan may have been influenced by the Chinese. Astronomy in ancient China was quite advanced, wasn't it?

IKEDA: Yes. The Chinese court always employed officials

whose main responsibility was to observe the stars and to make calendars.

SHIMURA: According to the *Shih chi* (*Records of the Historian*), the famous historian, Ssu-ma Ch'ien (145?-90? BC), of China held a hereditary position whose responsibility was to observe the movement of the stars and to keep records.

IKEDA: He had incited the wrath of the emperor and as a result was castrated. But he felt that he could not commit suicide and instead chose to endure this great humiliation so that he could complete the compilation of this work that he had undertaken. This provides a good example of one's given responsibility becoming a springboard for him to overcome human sufferings and to launch himself toward a greater mission. In ancient China, the emperor was regarded as the Lord of Heaven, who ruled the people on behalf of heaven. Information concerning the movement of the stars, based on the research done by a court official, was announced to the people by the emperor alone.

KIGUCHI: Predicting a solar eclipse must have been considered a golden opportunity for a ruler to display his power.

IKEDA: If I remember correctly, in the ancient Chinese book, *Shu ching* (*Classic of Documents*), there is a passage which draws a relationship between the emperor's authority and the occurrence of solar eclipses. I was told that based on the *Shu ching* which was written over two thousand years ago, modern-day astronomers were able to confirm when this eclipse actually occurred. More than four thousand years ago, in the Hsia period, there were two astronomers, I and He. They had neglected to inform the ruler of the next solar eclipse, and when it took place, people were taken by surprise. As a result, the emperor's authority was threatened. In anger, he had the two astronomers put to death.

KIGUCHI: If a source has the date and time of the solar eclipse, modern astronomers can determine the year.

Who Named the Constellations?

SHIMURA: The movement of the constellations has not changed much in the past four thousand years, has it?

KIGUCHI: It may appear that it has changed, but actually the way the stars move is very much the same. The constellations were devised by Chaldean shepherds in Mesopotamia, one of the cradles of ancient civilization, going back as far as five thousand years ago. While they herded their sheep, they watched the countless stars above, drawing imaginary lines between them, forming the shapes of people and animals. The result was twelve constellations, such as the bull, lion, scorpion and goat constellations.

IKEDA: Then these constellations were adopted by the Greeks in a fairly sophisticated form. I notice that there are no constellations depicting animals like elephant, alligators or tigers.

KIGUCHI: In Mesopotamia, several stones were excavated which were thought to have been used as stone markers. Depicted on them were the animals whose figures were represented in the constellations, along with the Sun, Moon and stars.

SHIMURA: At present, how many constellations are there?

KIGUCHI: The ancient Greeks categorized forty-eight constellations which could be observed in the northern hemisphere, but later the stars visible in the southern hemisphere were added. About fifty years ago, a meeting of the International Astronomical Union (IAU) brought the number of constellations up to eighty-eight, which are now used as references throughout the world.

IKEDA: Each country has its own designations of the constellations.

KIGUCHI: Yes. For example, Virgo is known as Virgin in the United Kingdom.

IKEDA: Isn't that confusing?

KIGUCHI: It can be; however, the official Latin designations are used by scientists throughout the world.

IKEDA: The night sky is especially magnificent in the summer time. Countless philosophers, heroes, rulers and poets, over the past five thousand years, have meditated and pondered on their dreams and the truth as they gazed up at the starry sky.

SHIMURA: In ancient times, after the fire died out, all they had was the light of the Moon and the stars.

IKEDA: When there was no electricity, people got up at sunrise and went to bed after sunset, underneath the glimmering light of the stars. It is said that many works of art were created out of the innate wisdom and sense of fulfilment that people derived from their close communion with the universe.

KIGUCHI: It goes without saying that people today have the ability to respond quickly to the latest scientific developments, but before such advances took place, people must have felt a greater sense of union with nature and the heavens than we can even imagine. There was also a sense of gratitude and reverence.

IKEDA: Based on this humble attitude, efforts to establish a relationship between human beings and the universe began to take placc.

KIGUCHI: We can say that space exploration, which is now the thrust of modern science, began, not with the desire for conquest, nor with the intention of using such technology for our own purposes, but with a kind of reverence.

SHIMURA: We are a lot further away from feeling the pulse of the universe. Moreover, in the large cities, we can no longer clearly see the stars at night.

IKEDA: About twenty years ago I visited the planetarium in Shibuya, Tokyo. I was surprised to find out that even in Tokyo, the sky above is like a jewel-box of glittering stars. Because of the Earth's rotation, the constellations seen in the evening gradually shift westward. As the eastern sky becomes brighter, totally different constellations appear. I was able to perceive the rotation of the Earth slowly and gracefully turning on its axis as I viewed the splendid pageant of the numerous stars that floated by. This experience is unforgettable, and I hope that all young people will have the same experience at some point in their lives.

SHIMURA: At the planetarium, one can witness solar eclipses, comets and even the orbits of man-made satellites.

KIGUCHI: It is said that the design of the planetarium originated in China as a derivative of the armillary sphere, an astronomical instrument.

IKEDA: There is one at the meteorological observatory in Beijing; it was made about two thousand years ago.

KIGUCHI: The ancient Chinese word *yuchou* (universe) has a very profound meaning.

IKEDA: The character *yu* indicates the limitless expanse of space, and *chou* signifies the eternal flow of time.

SHIMURA: I didn't realize that. But once we begin to turn our attention to the wonders of the universe, we can't help but feel awe and reverence.

IKEDA: Unless we relate to the universe and to other people with humility, we will never be able to understand them as they really are, let alone perceive the ultimate truth. The expanse of the universe is boundless, spreading out into the ten directions. The universe continues to exist throughout eternity. The 'ten directions', a term used in Buddhism, refers to the dimensions of space: north, south, east, west, northeast, northwest, southeast, southwest, up and down. Within this unlimited expanse, the unknown prevails. Although each phenomenon can be defined as existing, it cannot be understood with our intellect alone. Buddhism interprets the universe as being unfathomable. 'Unfathomable' is used to mean mystic, and the one Law which governs all the workings of the universe is called the Mystic Law.

In addition to the views of life and the universe as perceived on the basis of scientific analyses, there are other perspectives to be taken into consideration in connection with the Buddhist view of the universe, such as the idea of microcosm and life-space. Buddhism regards the universe as being filled with the potential for life and human life as a miniature universe. Further, it clarifies the relationship between the two, shedding light on the 'self', or the essence of life, which pulsates in harmony with the rhythm of the universe and permeates the infinite reaches of the cosmos.

KIGUCHI: The Buddhist concept of the universe itself as embodying 'life' is very far-sighted. In modern times, scientists view the universe within the framework of time and space, as indicated by the use of measurements and clocks. It wasn't until 1925 that the universe was considered from the standpoint of human consciousness. In cosmology, as practised today, with its emphasis on quantum mechanics, human subjectivity underlies our conceptions of the universe. Moreover, physicists boldly express their own ideas and personal conjectures about their research findings, which is quite a departure from twenty years ago. I believe that the study of space should include concepts like subjective- or life-space, an area of study which physicists might take an interest in.

IKEDA: It is important to understand the Buddhist concept of *kū*. *Kū* is one of the three truths, the other two being temporary existence and the Middle Way. The idea of *kū* differs according to the level of teachings, but here it means the state of being in its *latency* or *dormancy*. The concept of *kū* is also part of the four kalpas, the four-stage cycle of *jō* (formation), *jū* (continuance), *e* (decline) and *kū* (disintegration) which a world is said to repeatedly undergo. Buddhism expounds that all phenomena are subject to the principle of perpetual change. Moreover, it defines the ultimate reality underlying all phenomena as the 'self' which embodies both non-substantiality and temporary existence yet, in essence, is neither.

SHIMURA: The concept of *kū* must be very difficult for Western philosophy to grapple with. Later on, let us continue our discussion about *kū*.

The Feasibility of Substantiating the 'Grand Unified Theory'

KIGUCHI: Astronomy etymologically means 'the study of the stars', and its Japanese equivalent is *tenmongaku*, or 'the study of the heavenly patterns'. The movement of the heavenly bodies has so many kinds of patterns that ancient people felt that the universe had a will of its own. *Tenmongaku* can be regarded as means of deciphering these patterns.

SHIMURA: The English word 'cosmos' derives from the Greek word *kósmos* meaning 'order', 'world' or 'universe'. I think it is safe to surmise that in the Orient, the heavens are viewed from a more subjective standpoint, while in the Occident, they are viewed as physical phenomena to be reckoned with.

IKEDA: You may be right. The Greek word *kósmos* has the opposite meaning of 'chaos'.

KIGUCHI: That's right, and its inherent idea is that the world, which was born in chaos, gradually becomes less chaotic and then evolves into a state of 'cosmos'. In Buddhism, what concept would apply to this word?

IKEDA: The Buddhist teachings comprise innumerable sutras,

called the *eighty thousand teachings* to indicate a vast number. Among them, a system of thought pertaining to the one Law which governs the universe emerges from the standpoint of the so-called subjective life-space. Compared with Buddhism, the ancient Chinese or Greek concepts of the universe are partial and fragmentary. The Buddhist concept of 'the reality of the Law underlying and governing all phenomena in the entire expanse of the universe' embodies a more comprehensive life-philosophy than either the Chinese term *yuchou* or the Greek word *kósmos* can convey. It is my feeling that the Buddhist concept of universal phenomena will become all the more comprehensible when the 'grand unified field theory' posited by Albert Einstein is corroborated by modern science.

KIGUCHI: The so-called 'grand unified field theory' is as revolutionary as the unified field theory, which relates electro-magnetism and weak interaction, although it is not as yet completely proven.

The discovery of a new elementary particle, the 'Z^0 boson', was announced by the European Centre for Nuclear Research (CERN) at Geneva by on January 1983. It is regarded as a giant step toward proving the grand unified theory, because the new elementary particle relates the forces of electro-magnetism and weak interaction, substantiating the unified field theory.

The discovery of this elementary particle is indeed a great accomplishment on the part of European scientists. It is somewhat difficult to explain what it is in simple terms, but according to the grand unified theory, all forces which exist in the world, for example, gravity, electro-magnetic forces, and other strong and weak interactions exist inter-dependently and are derived from the same source of power originating at the time of the birth of the universe itself. This power is thought to manifest itself in various ways, according to environmental circumstances. If scientists can prove that there has always been only one source since the beginning of time, then the grand unified theory can be completely validated. It will mean the dawning of a new science.

IKEDA: In Buddhism it is said that numerous branches and

their leaves come from one root, and similarly, infinite meaning comes from the one Law. It took about two thousand years to clarify this root, or one Law – the Mystic Law.

KIGUCHI: If the grand unified theory is established, we will, scientifically, be another step closer to penetrating the depths of the Mystic Law.

IKEDA: The concept of the universe, as interpreted in the Chinese character *yuchou*, appears in Hinayana, or lesser vehicle, teachings. These teachings were employed as a means to lead people to the higher teachings and ultimately to enlightenment.

SHIMURA: According to Hinayana Buddhism, the world is conceptualized as having Mount Sumeru at the centre and the universe as consisting of innumerable world systems.

IKEDA: About fifteen hundred years ago, a Buddhist scolar from India, Vasubandhu, compiled the *Kusha ron* (Skt *Abhidharma-kosha-shāstra, A Treasury of Analyses of the Law*), after studying the teachings of Hinayana Buddhism.

SHIMURA: It is a guidebook to Buddhism.

IKEDA: It may be considered that. In the *Kusha ron* there is a chapter called 'Seken-hon' (On Realms), where the concept of the universe reflects that of ancient India. Vasubandhu was a monk thought to have lived around the fourth or fifth century AD. He had considerable knowledge and became the undisputed master of Hinayana teachings. Later, however, he was persuaded by his elder brother, Asanga, to convert to the Mahayana teachings. In recanting his error of adhering to the lesser teachings, he had tried to cut out his tongue. Thereafter, he greatly contributed to Mahayana Buddhism.

SHIMURA: Some scholars point out that the Buddhist concept of the universe is surprisingly similar to that of modern science. If this work, which is two thousand years old, were to be translated, we would certainly find this to be the case.

IKEDA: Yes, but in such works as the *Kusha ron* some of the passages are somewhat abstract.

SHIMURA: The other day, the Japanese physicist Dr. Kunitomo Sakurai said, 'Our contemporaries should become more appreciative of the glittering stars, which comprise nature's heavenly symphony.' Dr. Sakurai, at the invitation of the American Science Academy, had been long involved in research work at NASA.

IKEDA: I agree with his remark that it is more important to become intimate with the universe intuitively than to accumulate 'cold knowledge'. It is desirable to live a serene life with a broad perspective, communicating with the stars which fill the heavens.

'Pulsation' and 'Prayer' in the Universe

SHIMURA: People of aspiration can make their conceptions of the universe an integral part of their outlook on life and a source of nourishment for their daily lives.

IKEDA: An awareness of the pulsation of life which throbs, asserting its existence throughout the limitless expanse of the universe, no doubt enables one to elevate his life-state.

KIGUCHI: It's interesting that you use the word 'pulsation'. Some stars also pulsate repeatedly, first expanding, then contracting.

IKEDA: The other day I spoke with Dr. Carl Sagan's wife, who does research work with her husband. *Pioneer 10* is carrying a message to extraterrestrial beings, which includes a recording of the radio waves of a pulsar. She said that the pulsar sounded very much like a heart beat.

SHIMURA: A person who devotes his or her entire being to studying the universe can feel its lyricism or poetry.

IKEDA: Such dedication is not derived from cold rationale, void of emotion. Perhaps one can arrive at this understanding more fully if he lets his mind or consciousness expand infinitely beyond the frontiers of science. Buddhism refers to this kind of perception as 'the fire of the wisdom of enlightenment'. The profundity of the truth of the universe is much more than meets the eye, and is not limited to the Sun, Moon and the stars which we can observe. The Great Teacher T'ien-t'ai, or Chih-i (538-597), of China called this truth the 'region of the unfathomable'. The Great Teacher Dengyō, or Saichō (767-822), of Japan named it the 'ultimate truth free from all illusion'. The *Lotus Sutra* obliquely refers to this truth when it states, '... through the worlds in all ten directions, passage was unobstructed, as if they were one Buddha land.'

KIGUCHI: Although we cannot always perceive them with our senses, there are many forces exerting themselves on us, such as gravity and electro-magnetism. The atmospheric pressure alone equals the weight of five elephants on a man's back, which is about twenty tons.

IKEDA: Nichiren Daishōnin revealed the fundamental Law permeating all phenomena in the universe as the Mystic Law, and embodied it as the Gohonzon, or the true object of worship.

KIGUCHI: Then are the movements of the heavenly bodies due to the workings of the Mystic Law?

IKEDA: Nichiren Daishōnin wrote, 'You must realize that the revolution of the Sun and the Moon around the four continents is due to the power of the Buddhist Law.'

KIGUCHI: There is a limit to what scientific research and technology can accomplish, no matter how advanced they may become. I can understand the necessity of studying the teachings of Buddhism thoroughly.

SHIMURA: When you were in France in 1983, Mr. Ikeda, you said to the members there that *gongyō* is a ceremony of fusing one's life force with the power of the Gohonzon.

IKEDA: *Gongyō* is the basis of our Buddhist practice. It is called the 'ceremony to praise the Buddha's teaching'. The *Ongi kuden* states that to do *gongyō* is to eliminate discrimination. This means that under the True Law, all people are equal. When we observe *gongyō*, we sit before the Gohonzon, the embodiment of the true entity of all phenomena throughout the entire universe, and join our palms together. The act of joining one's palms together symbolizes the interaction and fusion of the individual's life with that of the universe. The *Fugen Sutra*, which concludes the *Lotus Sutra*, states that one should 'sit upright and meditate on the true entity'. This means that when we sit upright before the Gohonzon and single-mindedly pray to it, the true nature of the universe will manifest itself within our lives. Our ten fingers represent the mutual possession of the Ten Worlds. Two palms joined together symbolize the fusion of the objective reality, or truth of the Buddha nature inherent within one's life, and the subjective wisdom to realize that truth.

People Who Have Had Dialogues with the Universe

KIGUCHI: Regardless of the times or country, most people of accomplishment have had an interest in the cosmos to some extent. For example, there were Confucius, Chuang-tzu, Archimedes and Plato in ancient times and Kant, Goethe, da Vinci, Beethoven and Shakespeare in more recent times, to name a few.

IKEDA: In the Orient, there is an old saying, 'Polish your gem with a stone from another mountain.' This suggests that knowing others well will naturally lead to a better understanding of yourself. When a person faces an entity which is greater than himself, most notably the cosmos, he can experience a kind of awakening, sometimes within the 'flash' of an instant, and then his potential abilities come to the fore and blossom.

SHIMURA: Pasteur, the father of modern medicine, is a case in point. He was in the throes of attempting to disprove the theory of spontaneous generation. At that time he visited an astronomical observatory on a sudden impulse and there succeeded in carrying out his experiment.

KIGUCHI: This story is symbolic of the link between the macrocosm and the microcosm.

SHIMURA: American educator Dr. William S. Clark (1826-1886), famous for the saying, 'Boys, be ambitious!' left a great mark on Japan during the Meiji era (1868-1912), though he lectured at Sapporo Agricultural College (now Hokkaidō University) for a total of only eight months. He also devoted himself to studying the stars when he was young.

IKEDA: I heard that he became so involved in the study of meteorites that he finally took a doctorate in that field.

KIGUCHI: Is that so? At that time (1852), meteorites were not researched very much, so he must have been a vanguard in the field.

IKEDA: This was part of the spirit of Dr. Clark, a man who contributed greatly to the development of Hokkaidō. Also in that area, Takeaki Enomoto (1826-1908) stayed with his troops in the Goryōkaku in the Battle of Hakodate during the time of the Meiji Restoration (1868). He had an experience concerning meteorites, didn't he?

SHIMURA: Yes. When he became Japanese minister of the navy, he bought a meteorite which was being used as a weight for a pickle tub, and he had three long and short swords made from it.

IKEDA: I remember the swords being called 'shooting stars'. The first sword Enomoto saw made of meteorite was one which belonged to the Russian czar. This was during the time he served as the Japanese ambassador to Russia. He then studied how to make swords from a meteorite.

What We Learn from the Study of Meteorites

SHIMURA: A few years ago, when I visited Atsuta Village in Hokkaidō with the Japanese novelist Ryōtarō Shiba and others, I saw Enomoto's warship *Kaiyōmaru*, which was sunk in Esashi Harbour, being salvaged. At that time I heard that there are less than ten swords made from the materials of meteorites.

IKEDA: When I learned about meteorites in school at around the age of ten or so, I was very impressed. As soon as I went home, I looked the word up in a dictionary. I remember reading that meteorites are shooting stars which fall to the Earth before their mass is consumed in flames as a result of combustion which occurs upon entering the Earth's atmosphere. Is that correct?

KIGUCHI: That's a popular theory, but astronomers have not yet arrived at a final conclusion. A shooting star results when materials scattered in interplanetary space emit light caused by friction as they sail into the atmosphere at full speed, pulled in by the Earth's gravity. When this phenomenon occurs on a large scale, a meteor shower takes place.

IKEDA: In old paintings, shooting stars are depicted as being like a shower of rain.

KIGUCHI: This also happens when a comet is disintegrated by the Sun's tidal force, and small particles around its tail are caught in the Earth's orbit.

SHIMURA: Then are they also remnants of comets?

KIGUCHI: You can think of them that way. Moreover, they might be fragments of the many minor planets that are positioned between the orbits of Mars and Jupiter. In contrast, some types of meteorite, such as the one which fell at Allende in Mexico, are thought to contain the materials necessary to form planets.

IKEDA: Why can't we precisely identify the origin of meteorites even though we know for certain that they must have fallen from space?

KIGUCHI: Since we do not know the composition of those minor planets and comets, we cannot make any comparative analyses. Therefore, it is not possible to draw any conclusions about them.

IKEDA: About how many meteorites have been discovered on Earth so far?

KIGUCHI: I understand the number to be around 7,500 – 5,000 of which were collected in the Yamato mountains of the South Pole by observation parties from Japan and the United States.

IKEDA: It is said that analysis of meteorites yields a lot of data about the Earth's past and the birth of the solar system. Why is this?

KIGUCHI: The history of the solar system is, so to speak, encapsulated within meteorites. For instance, the percentage ratio of certain elements which make up a meteorite tells us to what extent the solar system was exposed to radioactivity in its early stages. In addition, its mineral components tell us about the temperature to which it was heated or the speed at which it was cooled. Meteorites are classified into three groups according to their mineral compositions: iron meteorite, stony-iron meteorite and stone meteorite.

SHIMURA: Most of the meteorites found so far are iron meteorites, such as the kind Enomoto used to make swords.

KIGUCHI: The largest meteorite in the world is the Hoba West Meteorite, which fell in south-west Africa. It is about three metres wide and weighs sixty tons. The materials which constitute this kind of meteorite were crystallized about 4.5 billion years ago.

IKEDA: That coincides with the age of the Earth.

KIGUCHI: It is believed that, roughly speaking, iron meteorites correspond to the Earth's core, stony-iron meteorites to its mantle, and stone meteorites to its crust.

IKEDA: So a mere stone can reveal to us a large chunk of history. Was this kind of study popular in ancient times?

KIGUCHI: No, it wasn't. Meteorites were generally exhibited in museums or, in Japan, sometimes became objects of worship in Shinto shrines. For instance, it is said that a meteorite weighing about one kilogram was found at a shrine near Nagoya.

SHIMURA: Not long ago, the United States announced that some meteorites collected at the South Pole came from the Moon.

KIGUCHI: American scientists have conducted a great deal of advanced research into stones and rocks from the Moon, and the announcement was the result of their preparatory studies, in which a similarity was noted between the properties of meteorites and those of the rocks from the Moon.

IKEDA: How many rocks were carried back from the Moon by the astronauts of the *Apollo* programme?

KIGUCHI: Around four hundred kilograms worth, I hear, and they have been distributed to thousands of scholars in over twenty countries.

IKEDA: A scholar once showed me a rock from the Moon. Do they reveal any new facts?

KIGUCHI: They show that the Moon was born around 4.5 billion years ago, and that its surface has holes ranging in size from one hundred to three hundred kilometres, made by meteorites which struck the Moon.

IKEDA: Then the debate about whether the Moon's craters were volcanoes or are the result of meteorites is settled?

KIGUCHI: Yes, the missions of *Apollo 15* and *Apollo 17* determined that they are the latter.

SHIMURA: How can we distinguish a meteorite from an ordinary stone or rock?

IKEDA: Meteorites have dents, just like ones we might make by pressing our fingers on clay.

KIGUCHI: They look like the dents on golf balls. The heat resulting from the friction with the Earth's atmosphere melts part of the surface to form the shallow dents. In addition, a layer about 0.3 millimetre thick forms all over the surface as the meteorite travels through the atmosphere. Occasionally one can see lines indicating the direction of the air flow. Moreover, when the gas, which later solidified within the Earth and meteorites, was still drifting in space, there were two stellar explosions nearby. This is believed to be the reason why meteorites contain radioactivity – though they are not harmful to humans after falling to the Earth.

SHIMURA: Is radioactivity emitted when a star explodes?

KIGUCHI: The explosion of a star is very much like that of a hydrogen bomb, so space is contaminated by the heavy metals. However, the elements of life were produced from matter which was radioactively contaminated.

IKEDA: So then we can say that atomic power is an integral part of the universe.

KIGUCHI: Yes, but we run the risk of disrupting the rhythm of the universe if we misuse atomic power.

SHIMURA: If we were to find a meteorite, what should we do?

KIGUCHI: Inform a science museum.

IKEDA: If we were to find a star that is not on the astronomical chart, what should we do?

KIGUCHI: Contact the Tokyo Astronomical Observatory as soon as possible. The finding will then be confirmed by a Schmidt telescope, which has a diameter of 105 centimetres, located at Kiso in Nagano Prefecture. After its confirmation, it is immediately reported to the Astronomical Telegraphic Office of the International Astronomical Union (IAU).

IKEDA: Is IAU the centre of astronomical studies throughout the world?

KIGUCHI: Yes. A new discovery is reported there, and if you are the first to have sighted it, your name is officially assigned to that star.

SHIMURA: Is this also the case with discovering a comet?

KIGUCHI: Yes, except that it is named after the first three people who sighted it, in the order that it was sighted. For example, a comet discovered by IRAS (Infra-red Astronomical Satellite) in April 1983 was later found by two amateur astronomers in Japan and the United Kingdom. So it was named IRAS-Araki-Alcock Comet.

The United Nations Flag, the Symbol of Mankind's Desire for Peace

SHIMURA: The concept of the universe has positive symbolic significance for both individuals and nations as a whole.

IKEDA: Time on a universal scale flows infinitely without interruption, while human beings and countries exist for a very short period of time. It is quite natural, therefore, that they look for a symbol of eternity.

KIGUCHI: This is clearly evident when we look at various national flags.

IKEDA: The stars shine eternally from on high, so the pursuit of national ideals is represented by stars on the flags.

SHIMURA: The United Nations Headquarters presented you with a United Nations flag in Zurich, Switzerland, in June 1983.

IKEDA: Yes, I received it in my capacity as president of the Soka Gakkai International (SGI). The background of the flag is pale blue, which symbolizes space, and in the middle, the Earth is depicted in white with the North Pole as its centre. Meridians are also drawn around the planet, and it is flanked by twin olive branches, as if to envelop the Earth in peace.

SHIMURA: It symbolizes that the United Nations is a world organization promoting peace.

IKEDA: That should be the mission of the United Nations. I have high regard for the United Nations' ideals, although we are a long way from realizing peace, and the organization has its problems and weaknesses. Nevertheless, the most effective way to achieve peace and order is for all countries and peoples in the world to support the United Nations.

KIGUCHI: The first cultural festival of the SGI was held at Trets in southern France.

IKEDA: Yes, and the UN flag was raised there, as it is at all of our international culture festivals. We also have the flags of the participating countries hoisted. I have noticed that many flags have designs with symbols of the Sun, Moon and the stars.

SHIMURA: The flag of the United States, the Stars and Stripes, is said to have been designed after the coat of arms of the first president, George Washington; it now has fifty stars. The flag of Burma has fourteen, the Chinese flag has five, and Iraq's, three. The Japanese flag bears the Sun, and the flags of Uruguay,

Malawi and Bangladesh also incorporate the Sun in their patterns.

KIGUCHI: The Islamic countries seem to prefer the crescent Moon.

SHIMURA: Countries located in the desert seem to find solace in the Moon rather than in the glaring Sun.

KIGUCHI: Perhaps, but man's relationship to the universe cannot always be explained in terms of physical phenomena.

SHIMURA: The human mind is indeed unfathomable.

IKEDA: As unfathomable as the universe itself. These flags seem to symbolize mankind's existence within the great scheme of the universe.

The 'Harmony of Life' Depicted in the Arts

SHIMURA: At the same time, people also want to become harmonized with the eternal flow of time and space.

IKEDA: People cannot always endure disorder and confusion; ultimately they long to live in harmony and unity.

KIGUCHI: That is a great truth; you mentioned this in the lecture you presented at the University of Bucharest in June 1983.

IKEDA: In that lecture, I spoke about the harmony of tradition and modernization. If we take examples from music and painting, this becomes all the more apparent.

KIGUCHI: How do you explain the harmonious relationship between the whole and the part?

IKEDA: This is difficult to explain. For instance, we cannot articulate why a painting moves us, no matter how much we might study it and closely examine the paint and canvas.

SHIMURA: The same thing can be said about music.

IKEDA: We don't enjoy beautiful music any better by analyzing the music score or the quality of the instruments. Great paintings and music created by master artists are possessed of a harmony which is universal. They capture the vivacious and exquisite harmony which exists within the human heart and within the universe.

SHIMURA: This is why such masterpieces and music move people, regardless of the age.

KIGUCHI: Kepler (1571-1630), the father of modern astronomy, wrote *Harmonies of the World*. As he speculated about the order of the universe, he meditated on the harmony of life. It was three hundred and fifty years ago when he liberated the study of the universe for the first time from the mysticism of Christianity, declaring, 'Astronomy is a study of physics.'

About Telescopes

IKEDA: Kepler's laws are famous, but he also devised a telescope.

KIGUCHI: He designed one which has a combination of convex lenses. With this device, heavenly objects were seen upside down.

IKEDA: On the lunar map which we see in photographs, I noticed that the top, which used to be labelled the south, became the north.

KIGUCHI: How did you come to notice this?

IKEDA: My third son, an elementary school teacher, has been

an astronomy enthusiast since childhood, so naturally I also became somewhat interested. One day, when I was comparing several pictures, I noticed the difference.

KIGUCHI: Since the *Apollo* spaceships landed on the Moon, the map of the Moon has been drawn in the same way that the maps of the Earth have been devised, according to the four directions – north, east, south and west.

IKEDA: I have some questions concerning telescopes. First of all, how different was Kepler's telescope from the one used by Galileo?

KIGUCHI: Galileo's was very simple. He ordered one which a Dutch optician had just invented and then adapted it to his needs. It had a convex lens for the object lens of the telescope and a concave lens for the eyepiece.

SHIMURA: I saw this telescope at a museum in Florence. It is one metre long, has the magnifying power of thirty, and is attached to a stand.

IKEDA: I also saw it once, and I was surprised that he was able to discover craters on the Moon, the rings of Saturn and the four large moons of Jupiter, all with such a simple telescope.

SHIMURA: He also discovered sunspots. It is said that he continually watched the Sun, even at the risk of losing his sight.

IKEDA: What size should a telescope be in order to be able to see the satellites of Jupiter?

KIGUCHI: If it has a lens of about ten centimetres, it should be sufficient.

SHIMURA: Would you talk a little about observing stars with a telescope?

KIGUCHI: Since the intensity of light fluctuates and is not

consistent, an image will be out of focus if the calibre of the telescope is insufficient. The same principle applies to the pupils of our eyes. The capability of distinguishing parts of an object is referred to as 'resolving power' or simply 'resolution'. With a telescope, the resolution's effectiveness is determined by the diameter of its lens. In the case of Jupiter, it is located about eight hundred million kilometres from the Earth, while the diameter of its satellite Io is four thousand kilometres, resulting in a ratio of two hundred thousand to one.

When we multiply this ratio by one twenty thousandth of a centimetre, the wavelength of light to which people are the most sensitive, the product is ten. Therefore, a telescope with a lens which is ten centimetres in diameter is the most effective.

IKEDA: What country manufactures the best telescopes in the world?

KIGUCHI: For amateur use, I think telescopes made in Japan are the best because the manufacturers such as Tōkyō Kōgaku and Nippon Kōgaku, researched the methods of making optical instruments exhaustively during the Second World War. When the United Kingdom's BBC did a story on Japanese astronomers some time ago, it conducted interviews with only the amateurs, completely overlooking the professionals. In the case of larger telescopes, which are used for serious astronomical study, a general comparison is impossible to make. Incidentally, the telescope at the Tokyo Astronomical Observatory is made by the Carl-Zeiss Foundation.

SHIMURA: What kinds of telescopes are now used to observe the celestial bodies?

KIGUCHI: The Earth has three different 'windows' which receive light of varying wavelengths from space. One is radio waves, another is infra-red rays, and the third is ordinary light. Accordingly, three kinds of telescopes are utilized: the radio-telescope, the infra-red telescope and the ordinary telescope.

IKEDA: Recently, telescopes carried on rockets and man-made

satellites which were sent up into space were used to conduct observations by means of ultra-violet, X-rays or gamma-rays.

KIGUCHI: Such observations become possible when there are no interferences, namely the atmosphere of the Earth itself. These days, the cosmos can be observed not only by electro-magnetic waves like light, but also by gravitational waves and neutrinos.

SHIMURA: A Japanese novelist, Yutaka Haniya, wrote a story in which, during the war, there were enforced blackouts along the Pacific coast of California because of the fear of Japanese air raids. Thus, the two-hundred-inch telescope at the Palomar Observatory was able to clearly observe many nebulae because of the darkness. Because of these discoveries, the study of astronomy really took off after the war, didn't it?

KIGUCHI: The results of observations depend on the prevailing conditions, and the blackouts at that time provided perfect opportunities. I understand that these days the lights from San Diego make observations difficult.

IKEDA: Many people think that all telescopes are cylindrical, but some are shaped like an antenna.

KIGUCHI: Aside from the cylindrical infra-red telescope, there are many telescopes of varying shapes. The radio-telescope is a parabolic antenna, while the gravitational wave telescope is a lump of aluminium or a single-crystal sapphire.

The neutrino telescope, now under construction in the deep sea off Hawaii, uses a layer of sea water, thousands of metres thick, as a dry plate. Another neutrino telescope, devised to observe nuclear fuel burning in the centre of the Sun, is made of carbon tetrachloride and weighs 610 tons. It is located 440 metres below the ground, deep within a gold mine of South Dakota.

IKEDA: Looking at photographs of the Sun, I am always surprised that the burning maculae, the corona and the gases which are gushing out can be clearly depicted, and that now we can even see the inside of the Sun clearly.

SHIMURA: Who took the first astronomical photograph?

KIGUCHI: About one hundred and fifty years ago, the American scientist, John W. Draper (1811-1882), is said to have taken photographs of the Moon for the first time.

IKEDA: Then the history of astronomical photography goes back a long way. If we were to put a camera in the back of a telescope, would we be able to take a picture of the Sun?

KIGUCHI: Yes, but because the light is very intense, we use a filter which allows only certain wavelengths to pass through.

SHIMURA: *Pioneer 10*, which is continuing its flight beyond our solar system, sent photographs of Altair to the Earth. You can learn a great deal from even these pictures, can't you?

KIGUCHI: Yes. By analyzing pictures of the surfaces of Mars and Venus, scientists have been able to determine the nature of the soil on those planets, and have now almost conclusively established that there are no living beings in existence there.

SHIMURA: A single picture can silence a hundred arguments.

IKEDA: Neither data nor logic can deny reality. Buddhism, therefore, attaches great importance to actual proof. In one of his writings, Nichiren Daishōnin teaches, 'Theoretical and documentary proof are important, but actual proof surpasses both.'

SHIMURA: These days, even physicians ultimately rely on photographs in determining their diagnoses.

KIGUCHI: In my field, no matter how plausible a theory may sound, it cannot be recognized as valid until it is put to the test and proven correct.

IKEDA: Buddhism employs the three proofs, documentary proof, theoretical proof and actual proof, as standards to decide the relative merits and depth of a religion. I think we can say that Mahayana Buddhism is rational.

The 'Five Types of Vision' in Buddhism

SHIMURA: As for the distances between the Earth and other celestial bodies, we can never perceive how far they are from us; we are simply overwhelmed by the sheer vastness of those distances.

IKEDA: They are certainly staggering to the mind, which is accustomed to distances that can be measured with a ruler or a tape measure. But after all is said and done, any unit of measure is nothing more than a product of human conceptualization.

KIGUCHI: We measure the distance between the Earth and a star, applying the same principle of triangulation used to make a map of a country or the globe.

SHIMURA: The distance between our two eyes is about six centimetres. How far can we see if we 'triangulate' with our eyes?

KIGUCHI: It is said that we can perceive distances of up to about ten kilometres.

IKEDA: So it's difficult but not impossible for a person to perceive the distance of ten kilometres.

KIGUCHI: The distances of stars are calculated at intervals of half a year, at the vernal and autumnal equinoxes for example, measuring the degrees by which they have moved in relation to the more distant stars. In this way, we can measure as far as about one hundred light-years.

SHIMURA: What about the more distant stars?

KIGUCHI: We calculate theoretically, based on the actual measurement of a nearby star. Of course, this isn't completely accurate.

IKEDA: Don't astronomers encounter problems if they make calculation errors?

KIGUCHI: We always devise theories that allow for margins of error. But returning to the subject of telescopes, it interests me that telescopes used to be called *tengan-kyō*, or 'divine-eye glass', in Japanese.

SHIMURA: *Tengan*, also pronounced as *tengen*, is a Buddhist term meaning 'divine eye'.

IKEDA: It is one of the five types of vision expounded in Buddhism. They are: the eye of common mortals, the divine eye, the eye of wisdom, the eye of the Law and the eye of the Buddha. As the saying goes, 'The eyes are the windows of the soul.' In all cases, to see implies that one's intellect, emotions and volition are functioning.

KIGUCHI: There is also an expression which goes, 'The eyes speak as much as the mouth.'

IKEDA: The eyes are expressive because one's emotions and will are reflected in them. Buddhism also teaches that the five types of vision denote five types of wisdom. For instance, the eye of common mortals means an ordinary human eye which distinguishes actual forms and phenomena as they appear to it.

KIGUCHI: For example, when we see the stars at night, we empirically recognize them as such. This is the perception of the eye of common mortals.

IKEDA: Yes, but in reality, the stars exist not only at night, but during the daytime as well. The ability to recognize this fact is said to be perception of the divine eye.

SHIMURA: One does not have to possess extensive knowledge in order to develop the divine eye. For instance, he doesn't have to be aware that light is a stream of particles, that we perceive brightness and colours according to the quantity of light that enters our eyes, or that we can't see the stars in the daytime because they are obscured by the sunlight.

IKEDA: The divine eye may be said to be the ability to perceive, by intuition, even events in distant places or phenomena not immediately discernible on the surface.

KIGUCHI: Some people acquire that ability through training.

IKEDA: There are others who possess this ability naturally, and they are sometimes said to have extra-sensory perception.

SHIMURA: Recently, I saw a series of pictures in a magazine of a yogi levitating.

IKEDA: In ancient times, one could be easily deceived by people claiming to have supernatural or occult powers. But now, in the age of rationalism and scientific discovery, such deceptions are easily revealed. Today, people desire to believe in a religion or philosophy that is based on the Law of life. Nichiren Daishōnin states that, in judging the validity of a religion, one should not be swayed by the supernatural or occult powers of its followers.

SHIMURA: Concerning the divine eye, there is the story of Maudgalyāyana and his mother, which was mentioned earlier.

IKEDA: That's a famous Buddhist parable. Maudgalyāyana was one of Shakyamuni Buddha's ten major disciples, known as the foremost in occult powers. His divine eye saw his dead mother suffering in the world of Hunger, and he tried to save her. But when he offered food to his starving mother, it burst into flames, and when he poured water over it, the water burned into firewood, making the fire bigger still.

SHIMURA: With his divine eye, Maudgalyāyana was able to see his mother's suffering and penetrate its cause.

IKEDA: He understood that, as karmic retribution for her greed and stinginess, his mother had fallen into the world of Hunger, destined to remain there for five hundred lifetimes. But, in spite of his divine eye and occult powers, he could not do anything to eliminate her suffering. Only after he had sought the Buddha's teaching and practised accordingly was he able to save his mother. This is said to be the origin of the *urabon* (Skt *ullambana*) ceremony.

KIGUCHI: It's perfectly understandable that in the past people called instruments used to observe things which they could not see with their naked eyes the 'divine-eye glass'. However, no matter how much in detail we might be able to perceive the phenomenal world through a telescope, we will not be able to discover its truth in its entirety.

SHIMURA: We must develop the wisdom to judge what we see with our divine eye.

IKEDA: This is the very function of the eye of wisdom. It enables one to penetrate deeply into phenomena, judge them correctly and understand the laws which govern them.

KIGUCHI: Which reminds me: though Kepler designed his own telescope, he could not observe the heavens himself because of his weak eyesight.

IKEDA: He had at his disposal all the data that his teacher, Tycho Brahe, had spent a lifetime collecting, and succeeded in formulating what are now called Kepler's laws with only a pen and paper.

KIGUCHI: These laws were what enabled Isaac Newton to discover the law of universal gravitation.

IKEDA: We might say that both these men were possessed of the eye of wisdom.

SHIMURA: Surely perception based on the eye of wisdom has contributed to the progress of civilization, although it has not necessarily added to human happiness.

IKEDA: The eye of wisdom alone cannot cultivate greater values and allow them to truly benefit mankind.

KIGUCHI: Human beings discovered that matter consists of nuclear energy, and succeeded in extracting it. It could have been harnessed for all kinds of beneficial purposes, but instead it was used for the first time in war. As a result, the world is now inundated with nuclear arms.

The Buddha's Eye: The 'Unity of "Self" and the Universe'

IKEDA: This is why the eye of the Law and the eye of the Buddha must be developed. The eye of the Law can also be called the eye of the bodhisattva. One who possesses the eye of the Law views everything and does everything from the standpoint of his determination as a Buddhist to strive for the salvation of all people. The highest respect for life and belief in absolute pacifism underlie all of his or her actions. The eye of the Law enables one to establish a way of life in which one always lives vigorously and cheerfully, as if he were taking a cruise on the sea of indestructible happiness.

KIGUCHI: Then the ultimate state is to develop the eye of the Buddha?

IKEDA: Yes. It is a state in which one realizes that the cosmos is the 'self' and vice versa. The *Juryō* chapter of the *Lotus Sutra* explains this as follows: '... the Tathāgata perceives the true aspect of the threefold world exactly as it is. There is no ebb and flow of birth and death, ...'

SHIMURA: What does this passage mean?

IKEDA: I'll explain it first from the standpoint of Shakyamuni's

Buddhism. The Tathāgata, that is, the Buddha, is a pure and powerful existence unswayed by anything. This magnanimous personality, with his peerless intuitive wisdom – that is, the eye of the Buddha – perceives that all things are entities of the ultimate Law of the Buddha, that is, the Mystic Law.

KIGUCHI: Then the eye of the Buddha represents what is referred to as enlightenment, the highest state that a person can achieve.

IKEDA: That can be one conclusion. In Shakyamuni's Buddhism, one could reach this state only after renouncing the mundane world and devoting oneself to Buddhist practice for a very long time. In marked contrast, Nichiren Daishōnin's Buddhism expounds that all people are inherently endowed with the eye of the Buddha as well as the four other types of vision. One is therefore able to attain enlightenment through faith in the Mystic Law. This is referred to as 'immediate attainment of the ultimate truth' or 'embracing the Mystic Law is in itself enlightenment'.

SHIMURA: What does 'the true aspect of the threefold world' mean?

IKEDA: The threefold world indicates the world of desire, the world of form and the world of formlessness. To perceive the true aspect of the threefold world means to realize that it is filled with anguish and illusion. In other words, it is a world where people must undergo the four sufferings of birth, old age, sickness and death, a world subject to the four kalpas of formation, continuance, decline and disintegration and to the four aspects of appearance, continuance, change and disappearance. It is a world in which all things and phenomena are impermanent, transient and in a state of constant flux. The Buddha's eye, however, perceives that, underlying all such changes and motions, there exists a permanent, unchangeable Law, that is, the Mystic Law. It is always aware of a world of eternally tranquil light pulsating in a steady rhythm in the depths of all things and phenomena, in society, on Earth and throughout the universe.

KIGUCHI: That is a very profound concept. Could you explain the phrase, 'there is no ebb and flow of birth and death'?

IKEDA: That is the essence of the Buddhist philosophy of life, the core of Buddhist cosmology. In brief, it means that, by penetrating the Law of the universe and acting in accordance with it, one enters into a tranquil state in which he perceives the true, eternal entity of life, absolutely unfettered by man's most vital problem, which is the issue of life and death.

KIGUCHI: It follows, then, that the principles of Buddhism operate in the very act of living, within the dynamic realities of society. We can develop and bring to perfection the eye of the Law and the eye of the Buddha as we lead our lives from day to day.

IKEDA: That's right. Nichiren Daishōnin states, 'The five types of vision are born of the *Lotus Sutra*,' the *Lotus Sutra* here denoting the Mystic Law. As we embrace and practise the Mystic Law, striving for our human revolution, we will naturally acquire the power, wisdom and vision to create peace, happiness, a harmonious society and a fulfilled life.

VI

Approaching Death's True Aspect from the Standpoint of Buddhism

Death concludes one's life in this world; exploring its hidden realm is the surest path to the establishment of the sanctity of life and, further, to lasting peace on Earth.

The Buddhist Concept of Death

SHIMURA: The problem of life and death is the most fundamental one for human beings.

IKEDA: Unless we come to terms with this vital issue, whatever we might achieve in the way of wealth or fame will ultimately be without purpose.

SHIMURA: I have witnessed the deaths of several individuals, and observed that most of them, as they faced death, experienced torment. Besides, there are many ways to die. John F. Kennedy was assassinated, Adolf Hitler committed suicide, supposedly with his wife, Eva Braun. General Maresuke Nogi, a hero during the Russo-Japanese War (1904-1905), and his wife killed themselves after the death of their lord, the Emperor Meiji, and Hideki Tōjō, the premier who led Japan to war, was declared a war criminal by the International Military Tribunal and executed.

IKEDA: Some deaths result from sickness, suicide, murder, accidents and euthanasia. Other so-called unnatural deaths also include death by drowning, execution, fire, etc. A natural death may be said to be the end of a natural life span when one loses consciousness and falls into a state resembling peaceful sleep.

SHIMURA: How are these differences in the way one dies interpreted in Buddhism?

IKEDA: In Buddhism, death is sometimes put in the category of *shima* or devil of death, a figurative expression depicting the serious impact of death on one's life, which is listed among the three obstacles and four devils as clarified in Buddhism. Death is also the result of karmic retribution or immutable karma. With these and other explanations Buddhism serves to emphasize the cause-and effect relationship involved in death.

KIGUCHI: In other words, Buddhism's concept of death is plausible to people in that it is explained in terms of the law of causality.

IKEDA: The many different causes and types of death reflect the many kinds of latent effects[1] within people's lives. Death does not mean the annihilation of life or the extinguishing of karma. From ancient times to the present, in both East and West, philosophy and religion have pondered the question of life and death. No doubt this is because of the feasibility that life may exist, in some form or other, after death, and it is a possibility that must not be overlooked. Christianity expounds the concept of an everlasting soul. The late Austrian political philosopher Count Coudenhove-Kalergi said that, beginning with Pythagoras and Plato, great philosophers in Europe have believed in life after death.

The differences that arise from within each individual life can be said to be the manifestation of latent effects which derive from internal causes made in past lives. These differences include those of personal character, social standing, physical characteristics, etc. Why some people are born in a particular country, have good health or bad health, or die young or at a ripe old age are questions whose answers are open to conjecture.

KIGUCHI: Such questions go beyond the realm of science.

SHIMURA: The mysteries of life defy explanation because it

cannot be measured or understood on the basis of logic or reason alone.

IKEDA: Buddhism expounds the Law which governs life and reveals that the life force of each individual is intrinsically linked with the life of the entire universe.

Suicide: an Act That Transgresses the Law

SHIMURA: Recently, there has been an increase in the suicide rate. It would appear to be a worldwide problem, and what's even more distressing is the mounting number of family suicides in Japan.

IKEDA: Even the best efforts on the part of government, science or education have not been able to alleviate this growing trend. It is indeed a tragedy. This phenomenon reflects the lives of those individuals who are bound by the irresistible power of karma, having no means to escape it. During the Heian period (794-1185) in Japan, for example, there was virtually no suicide. Perhaps it was because the Japanese cultural climate at that time was positive. If one experienced a feeling of emptiness or encountered a stalemate in life, one left his home to pursue a religious life, seeking the Way of Buddhism.

SHIMURA: It would seem that the suicide rate is increasing along with the growing complexity of society.

IKEDA: It is most unfortunate that so many well-known figures resorted to suicide. In particular, I'm thinking of the Japanese author, Yukio Mishima, who killed himself in 1970, and Yasunari Kawabata, another Japanese writer who committed suicide in 1968, four years after receiving the Nobel Prize for literature. In other countries, there was Ernest Hemingway, 1954 Nobel Prize-winning novelist who killed himself after becoming mentally deranged, Lao She who reportedly committed suicide during the Cultural Revolution in China, and just recently, the British author, Arthur Koestler.

At one point, I was to have met Kawabata, but unfortunately it didn't work out. However, I did meet Mishima at the Hotel Ōkura in Tokyo. He was a friend of Hiroshi Hōjō, fourth president of the Sōka Gakkai. They met during the time that both of them attended Peers School. The death of these two great authors is a tremendous loss. When I visited Romania in June of 1983 and met members of the writers' union, I discovered that the names of Kawabata and Mishima were very familiar to them. Their books were read in translation. Moreover *The Tale of Genji* and the works of Jun'ichirō Tanizaki were discussed, and they had also studied *haiku* enthusiastically, particularly those of Bashō and Buson.

SHIMURA: Arthur Koestler's death caused a stir in France.

IKEDA: René Huyghe told me about the suicide with such a serious coutenance. His was the most solemn expression that I had ever seen.

SHIMURA: Koestler wrote *Darkness at Noon* (1970), and in his later years he was greatly involved in issues concerning life and the universe. He committed suicide right after the publication of his book *Janus* (1983). It's a great pity.

IKEDA: During my talks with Dr. Toynbee, I asked what had been the most tragic moment of his life, and he replied with the most grieved expression on his face that it was when his son committed suicide in an adjoining room by shooting himself with a pistol. I don't think that this was a matter he discussed with many people.

SHIMURA: The book *Suicide mode d'emploi* written by Claude Guillon and Yves Le Bonniéc, published in France in 1982, was translated into Japanese recently and has been receiving a lot of attention. What kind of mental anguish motivates people to commit suicide?

IKEDA: With a sense of complete hopelessness, people sometimes kill themselves out of a desire to escape. Buddhism teaches that the human being is the vessel of the Law or the Dharma. To violate one's own body or rob this vessel of precious life is considered to equal the crime of 'theft'.

Buddhism's Position on Capital Punishment

KIGUCHI: There is now a raging controversy taking place on an almost worldwide scale concerning the death penalty. What are your thoughts on this, Mr. Ikeda?

IKEDA: I am against capital punishment. There is a story concerning King Ashoka when Buddhism was flourishing in India in the third century BC. He was the first to unify nearly all of India, and was loved because of his benevolent rule based on Buddhist mercy. He abolished capital punishment, had a lenient attitude toward offenders and worked to promote friendly relations and exchanges with other countries. Moreover, he put much energy into furthering culture. As a Buddhist, he espoused absolute pacifism.

During the Kamakura period (1185-1333), prior to the second invasion of Japan by the Mongols,[2] the feudal government of Japan beheaded Mongol delegates. Nichiren Daishōnin reacted strongly and condemned the government, declaring, 'That innocent envoys should be beheaded is most tragic.'

Buddhism teaches people how to create the most valuable and meaningful existence. Enlightenment is the ultimate goal of a human being's existence, and famous passages in the sutras talk about 'complete fulfilment' and state that, because of Buddhist practice, 'the people are happy and at ease.' It is of the greatest importance to lead a happy and fulfilled life as a human being.

SHIMURA: You mentioned earlier that when you were in Romania, *The Tale of Genji* was a topic of discussion. What can we learn about death from this book?

IKEDA: The four sufferings of birth, sickness, old age and death are woven through the entire book. Written during the

Heian period by Murasaki Shikibu (c. 1000), it is considered to be one of the great works of Japanese literature. Four metaphors are used to convey the mood or appearance of death. The author, through the protagonist, Hikaru Genji, used the expressions, 'to vanish as foam', 'like the dew about to vanish away', 'to be vanishing, fading away like a flower' and 'like a dying flame'.[3]

Shikibu often used autumn as the setting whenever a death occurred in the story, although several women died in the spring perhaps because the author, a woman, felt a kind of empathy. The writer may have used the metaphor of death in the spring to signify an earlier reincarnation. At any rate, according to the Buddhist view of death, one's final reckoning takes place at the moment of one's death.

If it weren't for this reality, people would no doubt lapse into hedonism, living only for the moment. There would be no progress. As a result, human existence would be devoid of any sense of responsibility. The inevitability of one's final reckoning is the main motivation for disciplining oneself, performing good works for others and contributing to society.

SHIMURA: This is indeed the case.

IKEDA: When Nichiren Daishōnin talks about the 'strict Law of the three thousand realms', he is referring to the Law which is inherent within one's life at every moment, and in accordance with this Law every cause one makes accumulates and becomes part of his karma as he transmigrates from one existence to another. In his writing, *Nyonin jōbutsu shō* (*Women's Attainment of Buddhahood*), the Daishōnin states, 'According to the sutras, a person experiences over 800,004,000 states of mind in a single day.' The corresponding causes of each of these momentary life-states are said to accrue in one's life and become part of one's stored karma.

SHIMURA: If all causes were to dissipate in obscurity, one would not have a seeking mind to learn and understand anything about life. Moreover, one might be content to lead a life as a bystander, merely observing society and possibly even his own life. However, there is no escape from cause and effect, regardless of how unrelated events may appear on the surface.

IKEDA: Nichiren Daishōnin also said, 'First understand death, and then go on to learn everything else.' The 'everything else' he refers to could be politics, economics and so forth. These concerns are indeed important, but our first priority should be placed on resolving the fundamental problem of life and death – this should be the human being's greatest challenge. Today, a great emphasis is placed on politics, science and economics. While approaching problems from these standpoints may appear to be the best way to solve them, I feel that people's priorities have been reversed.

SHIMURA: Perhaps this is because human beings are unable to maintain a constant awareness of the reality that they will eventually die. Is there any way that we can know when we will die?

IKEDA: Ordinary common mortals cannot predict the time of their death. However, just before death, some people who are seriously ill can sense that their time is limited. Many doctors have observed that when patients say things like, 'It won't be long now,' or 'Thanks for everything you've done for me,' they usually die within the next twenty-four hours. Of course, we cannot predict exactly the year, month, day or time of death.

KIGUCHI: What about the Buddha?

IKEDA: I would like to refer you to the teaching of Nichikan Shōnin, the twenty-sixth high priest of Nichiren Shōshū: 'Shakyamuni Buddha, after expounding the teaching of the wonderful Law on Eagle Peak, entered nirvana in the grove of sal trees by the Hiranyavatī River, located to the northeast of the mountain. [In the same manner,] Nichiren Daishōnin, after expounding and chanting the Mystic Law on Mount Minobu, passed away at Ikegami Village by the Tama River, located to the northeast of the mountain.' We can see a similarity in the way that both Shakyamuni and the Daishōnin passed away, in a fashion that we ourselves cannot contemplate imitating.

In perceiving what the future holds for one, the Daishōnin says, 'A sage is one who fully understands the three existences of life – past, present and future,' and 'With such perception one can see the distant future by looking at what is close at hand. One can infer what will be from what exists in the present.' This ability to perceive the time of their death is enough to prove their credibility. This is why we can believe in the teachings of these great sages.

The *Lotus Sutra* says, '... the Tathāgata perceives the true aspect of the threefold world exactly as it is.' Buddhism, which is based on the enlightenment of the Buddha, enables its followers to perceive the true aspect of all phenomena or the true entity of life. When we contrast this teaching with, say, the philosophy of Marx, Hegel or Kant, which has its origins in Greek thought, and which deals with phenomena within the realm of human society based on reason or knowledge, we can clearly see the difference.

SHIMURA: Western philosophy has made great contributions in terms of the improvement of social conditions, the development of an established historical perspective and the advancement of science; these endeavours are within the confines of the objectivity of human affairs and civilization. However, human beings, possessed of an internal or spiritual side which is at best enigmatic, cannot grasp the entire reality of their existence on the basis of objective observation only.

IKEDA: In trying to understand life itself, people must somehow also develop a subjective approach; in one's search for the absolute truth, doing so must inevitably lead to religion and faith.

KIGUCHI: In other words, when we view the human being *subjectively*, we can come to understand that part of life which is intangible, and when we observe the human being *objectively*, we can discern the dynamics of life within the context of social activities and daily existence.

IKEDA: Mr. Toda made a very interesting remark. He said, 'If we knew what year and day we were to die, our existence would be one of pain and anguish. Perhaps it is the Buddha's mercy that we don't know; as common mortals we are much better off and happier by not knowing.' In any case, the cycle of life and death is eternal. This is a fundamental principle of Buddhism.

In the *Juryō* (sixteenth) chapter of the *Lotus Sutra*, the heart of the so-called eighty thousand teachings of Buddhism, there is a place which reads, 'I [the Buddha] let the people witness my nirvana as a means to save them.' It interprets death as a 'means' to achieve some purpose. Let's compare a lifetime to a day, for the sake of explanation. In the morning we get up, which is a kind of 'birth' and the beginning of the day's activities. At night we go to sleep so that we can become refreshed for the next day's 'rebirth'. Resting is a kind of 'death', the ending of one's life for the day. Similarly, after completing a life of fulfilment, we die. Death is an expediency which enables us to renew our life-energy and to be reborn after having accumulated vital life force. Life, which spans the life-conditions of the Ten Worlds, undergoes this transmigration repeatedly, for eternity. This is one's true aspect of life, and I believe that this is what the passage in the *Lotus Sutra* actually means.

The Life Span of the Sun, Moon and Earth

SHIMURA: We now know that stars are also born and then die; can you tell us, Dr. Kiguchi, anything about their span of existence?

KIGUCHI: Stars have their respective life spans, but as I mentioned previously, stars are formed from gaseous clouds. The size of the star's mass at the time it condenses after cooling is what determines the span of its existence.

IKEDA: I understand that the bigger the mass of the star, the shorter its life span. Is this because, due to its increased motion and greater activity, its energy is expended that much more quickly?

KIGUCHI: That's right. On the other hand, stars with less mass have longer life spans because they don't consume as much energy.

SHIMURA: What is the estimated life span of the Earth?

KIGUCHI: The Earth is calculated to be 4.5 billion years old now, but the remaining years left to it are closely linked to its relationship with the Sun.

IKEDA: I heard that within about five to six billion years, the Sun's energy will be exhausted and it will begin to collapse.

KIGUCHI: The Sun is about 4.5 billion years old now; so accordingly, its life span is estimated to be about ten billion years.

SHIMURA: How are those figures calculated?

KIGUCHI: Putting it simply, during the fusion process in the Sun, hydrogen is converted into helium and energy, which are released in the form of light. We can calculate how much heat the Sun radiates during a specific unit of time, taking into account the distance between the Earth and the Sun and the amount of heat which the Earth receives. Then, given the Sun's weight, we can calculate the volume of combustible fuel, and from these figures we can estimate the Sun's life span.

SHIMURA: What about the Earth?

KIGUCHI: Our estimations are based on the amount of specific radio-isotopes contained in the Earth's minerals.

SHIMURA: And the Moon?

KIGUCHI: We haven't fully analyzed it yet, but we know that it came into being at around the same time as the Earth did.

IKEDA: The Sun is about 4.5 billion years, and so are the

Earth and Moon. This indicates that there is a definite relationship among these three bodies.

SHIMURA: What about other heavenly bodies?

KIGUCHI: Unfortunately, there is an insufficient amount of data from other planets in our solar system, but most of them are believed to be about the same age as the Earth. When the Sun begins to deteriorate and the amount of available hydrogen decreases, the Sun will start to expand.

IKEDA: When this happens, traumatic changes will occur on Earth. Temperatures will soar and the ecological balance will collapse; everything will burn up completely.

KIGUCHI: The Sun will swell to one hundred to two hundred times its present size, its intense heat engulfing Mercury, Venus and the Earth. A star in this phase is called a 'red giant'. We can actually observe other stars at this stage. The Sun will continue to expand for two billion years, and then suddenly it will contract. A colossal explosion will take place, and the outer layers of the Sun will be blown about in the form of gases and dust. This is what is known as a planetary nebula. The remaining part of the Sun will become a very small hot star known as a 'white dwarf'.

IKEDA: That's amazing. The death of the Sun is something so great and dynamic as to defy the limits of our imagination. But, seen in the light of Buddhism, even the movements of the Sun, Moon and Earth all follow the Buddhist principle of the four kalpas or the four stages in the cycle of formation, continuance, decline and disintegration, as is expounded in the sutras. As we discussed earlier, when we consider the universe as a whole, it is presently expanding. The Mystic Law permeates all phenomena throughout the infinite universe. On our planet, we can see this Law operating within people's lives. Moreover, faith in this Law and the philosophy of human happiness and lasting peace is now spreading throughout the entire world. In any event, all things in the universe follow the cycle of formation, continuance, decline and disintegration. This is further proof of the validity and profundity of the Buddhist teachings.

In terms of the four kalpas, the existence of human life can be understood in the following way: the Kalpa of Formation would be equivalent to one's birth and youth; the Kalpa of Continuance, to one's prime of life; the Kalpa of Decline, to one's old age; and the Kalpa of Disintegration, to one's death.

This idea could also apply to any given day: formation indicates the morning; continuance, the afternoon; decline, the evening; and disintegration, the stillness of the night. In terms of a year, formation would be spring; continuance, summer; decline, autumn; and disintegration, winter.

In order to grasp the wisdom of Buddhism more clearly when making a comparison with Western philosophy, we must probe deeper and discuss the concept of existence.

At the risk of being somewhat abstruse, we must bring up the Buddhist concept of *kū* or *non-substantiality*, which is sometimes also translated as *void* or *latency*. This is the core of Mahayana doctrine. It posits an elusive reality which transcends both the words and concepts of existence and non-existence. In Buddhist doctrine, this is said to be the one and only truth of the Middle Way, or the 'self'. This 'self' becomes manifest within the form of a living being as a result of a combination of various causes and relationships; its birth is equivalent to formation. This phenomenon is in keeping with the universal rhythm which constantly undergoes the cycle of the 'four kalpas'.

SHIMURA: It seems that a number of astronomers are now realizing that there is a similarity between the Buddhist concept of the Law which governs the cycle of the birth and death of stars, and present scientific findings. Recently, even an editorial in Japan's leading economic journal *Nihon Keizai Shinbun* observed that modern science's concept of the universe is very similar to that of the 'four kalpas'.

KIGUCHI: There are indeed many assumptions about universal phenomena which are shared by Buddhism and by the discoveries of modern astronomy. Although the birth and death of stars and the law of gravitation may appear to be just physical phenomena, in order to grasp their true aspect we need to observe these occurrences from a subjective viewpoint as well.

SHIMURA: The German physicist Werner Heisenberg (1901-1976), said to be one of the greatest scientists of this century, had a similar idea.

IKEDA: He is particulary well known for his so-called 'indeterminacy principle', but after he received the Nobel Prize, Heisenberg leaned more towards believing that there exists some kind of order which pervades all natural phenomena.

KIGUCHI: The indeterminacy principle ushered in a new epoch in twentieth-century scientific research, completely different from previous achievements, including the development of Newtonian mechanics and Charles Darwin's theory of evolution. Moreover, we can also say that Heisenberg's revolutionary theory has restored the importance of man's subjective thinking in the development of the physical sciences.

IKEDA: For example, Heisenberg noted that what we observe with our scientific eye is not the true nature of the phenomenon under observation. Our understanding is restricted by the method of observation which captures only part of the picture.

KIGUCHI: Heisenberg and the Danish physicist Niels Bohr (1885-1962) were both aware of the inseparable relationship between what is observed and the observer – between the object and the subject.

Dr. Bohr was a renowned Nobel Prize winner, who also taught many other prominent scholars around the world. The view of both these scientists is referred to as the 'Copenhagen interpretation' among quantum physicists, and this led to a new wave of development within Western philosophy during the first half of the twentieth century. The British philosopher Alfred Whitehead (1861-1947), known for his thoughts on meta physics, and logical positivists such as Hans Reichenbach (1891-1953) and Rudolph Carnap (1891-1970) were strongly influenced by the 'Copenhagen interpretation'. However, even this proved to be unsatisfactory to many physicists.

SHIMURA: It's a well-known fact that both Bohr and Heisenberg came to have great respect for Buddhism later on in their lives.

IKEDA: It seems that a great many scientists tend to be drawn to religion sooner or later. Interestingly enough, many such scientists have been drawn to Buddhist teachings.

In Buddhism, there is a concept referred to as 'the oneness of life (subject) and its environment (object)'. 'Oneness' here means 'two but not two, not two but two'. Life and its environment, while they are considered two separate phenomena, are actually one and inseparable in their essence. Nichiren Daishōnin says in the *Zuisō gosho* (*On Omens*), 'All that is in the ten directions of the universe is "environment", while all living things are "life". If we say that the "environment" is like a shadow, then "life" is likened to the body which casts its shadow. Without the body there can be no shadow. Similarly, without life, the environment cannot exist, even though life is supported by its environment.' This clearly explains the inseparability of human life and its surroundings.

KIGUCHI: In scientific research, living beings and the environment had, for a long time, been conceived of as two separate phenomena. Science made great strides, but as it progressed, it also went up a few blind alleys.

Halley's Comet

SHIMURA: To change the subject, I hear that Halley's Comet will soon appear after a seventy-six-year interval.

KIGUCHI: It is presently heading in the direction of the Sun on its way back, on its predictable course and on schedule.

IKEDA: The returning comet was seen from the Palomar Observatory in the United States on 16 October 1982, but it will take three more years before it is visible to the naked eye.

KIGUCHI: Yes, Halley's Comet could be seen nearing the orbit of Saturn; it registered a brightness of a magnitude 24.

SHIMURA: In terms of measurement, what is a 'magnitude' and how far does the scale go?

IKEDA: I remember reading somewhere that a person with good eyesight can perceive on a clear night an object of magnitude 6.

KIGUCHI: And stars a hundred times brighter are considered to be of the first magnitude. If they are any brighter, we add a minus. This seems contradictory, but the weaker a star's brightness, the larger the star's magnitude number.

IKEDA: Which means that for us, Venus has one of the highest magnitudes, which is −4.4. But Venus doesn't seem so bright in Tokyo, does it?

KIGUCHI: That's because of the city lights and the smog.

SHIMURA: How is the magnitude of a star determined?

KIGUCHI: The magnitude is decided upon when a star, at its brightest intensity, is sighted from the Earth. Strictly speaking, when we refer to the magnitude of a star, we mean its greatest magnitude.

SHIMURA: Can all the planets in our solar system be seen with the naked eye?

KIGUCHI: It is impossible to see Neptune and Pluto.

IKEDA: And Saturn?

KIGUCHI: It has a magnitude of −0.4, which makes it visible in principle. Mars, which is the brightest planet after Venus, has a magnitude of −2.8.

IKEDA: What is the magnitude of the Sun?

KIGUCHI: About −26.

IKEDA: And the Moon?

KIGUCHI: The Moon, at full phase, is magnitude −12.

SHIMURA: How can Halley's Comet, with a magnitude of 24, be detected?

KIGUCHI: It was sighted using both a five-metre (200-inch) reflecting telescope and sophisticated electronic technology.

IKEDA: That's a tremendous achievement. Where is the comet now?

KIGUCHI: According to the calculations made by the Palomar Observatory, it will be in the vicinity of the orbit of Saturn until the end of August 1983. In September 1983 it will travel through Saturn's orbit, and during all of 1984 it will be on the far side of Jupiter's orbit.

IKEDA: I understand that its speed increases as it approaches the Sun.

KIGUCHI: The force of the Sun's gravitation causes the increase in speed. Around November 1984, it will enter Mars's orbit.

IKEDA: And then from there, because of the Sun's heat, the famous tail of the comet will stream behind it.

KIGUCHI: The head of the comet is called the nucleus. This is a mass of frozen carbon dioxide and cosmic dust. The head absorbs heat, and a part of the head vaporizes and gases surround the nucleus. This form is the so-called *coma*, and as the comet approaches the Sun, the length of its tail increases. Eventually, it looks as if it has long hair trailing behind it.

IKEDA: I heard that the English word comet has its origin in the Greek word *komētēs*. Is that correct?

SHIMURA: Yes, and it means *long-haired*.

KIGUCHI: The closer the comet approaches the Sun, the longer its tail becomes. When the comet is at the closest point to the Sun, which is called *perihelion*, the tail is at its longest. In 1910, it is recorded that there was widespread panic because of the rumour that the tail contained poisonous gases, and that all living beings exposed to it would surely die.

IKEDA: Someone's remarks aroused the people's fear which was based on their lack of scientific knowledge. I understand also that there is a connection between the appearance of Halley's Comet and the sunspot cycle.

KIGUCHI: There are variations in the number of sunspots and in other aspects of the Sun. We have been able to determine an eleven-year cycle, but there may also be an eighty-year cycle, which roughly coincides with the appearance of Halley's Comet.

SHIMURA: To what extent can this connection be scientifically verified?

KIGUCHI: There isn't enough data at present. With reference to the movement of celestial bodies, there are still too many phenomena yet to be explained. Even when scientists can explain 'how', they are far from being able to explain 'why' in many cases.

IKEDA: There are phenomena in the universe which the realm

of science can clarify, but there are others which are beyond scientific description. Of course, research will enable us to gain a better understanding, but at the same time, we must deal with phenomena which cannot be explained in the light of present scientific research. Buddhism is a philosophy which clarifies the distinction between the two.

KIGUCHI: What scientists must acknowledge are the limitations of science, although the constant effort to make further progress is of course important.

SHIMURA: When will we be able to see Halley's Comet from the Earth?

KIGUCHI: In November of 1985 and again in the following April. However, in April the position of the comet will be such that it cannot be easily observed from the Earth. Nevertheless, those of us in Tokyo who get up early in the morning, about 4.00 am on 2 April 1986, should be able to see it in the lower part of the southern sky.

IKEDA: I hear that Japan is planning to observe Halley's Comet using artificial satellites.

KIGUCHI: Yes. We'll launch an interplanetary explorer in August of 1985. But we're not only going to view the comet at close range; we'll actually try to extract some materials for analysis.

Phenomena in the Heavens Recorded in Ancient Chronicles

SHIMURA: There are a number of records which mention the appearances of Halley's Comet in ancient chronicles all over the world.

IKEDA: In Japan, there are well-known references in the documents of the Kamakura period. Among them is the *Azuma kagami* (*Mirror of Eastern Japan*), a historical record of the Kamakura shogunate, which is known widely.

KIGUCHI: How is it recorded?

IKEDA: For example, on the second day of the eighth month of the first year of Jōō (1222) it is reported that there appeared a comet half the size of the Moon, with a length of about 5.5 metres. Later on in history, it was confirmed that this was Halley's Comet during its earlier stage. In China, the brilliance of the comet so shocked the people that they changed the name of the era. In Korea, it is recorded that the comet was visible even during daylight.

SHIMURA: It's fascinating to compare the astronomical data with these historical documents.

KIGUCHI: This particular field of study is called archaeo-astronomy. As far as I know, there is only one report of Halley's Comet being visible during daylight hours.

IKEDA: It's interesting to me that this should have occurred during the same year that Nichiren Daishōnin was born.

SHIMURA: Speaking of documented observations, a new comet, sighted on 3 May 1983, missed the Earth by a very close margin.

KIGUCHI: Yes, that was the IRAS-Araki-Alcock Comet. The near-miss occurred on 12 May, and even astronomers were a little shaken, although the comet disappeared within ten days of its discovery.

SHIMURA: How close did it come?

KIGUCHI: About 4.5 million kilometres.

IKEDA: Before that one, what was the closest approach?

KIGUCHI: In 1770, Lexell's Comet came within 3.2 million kilometres, making the one which occurred in 1983 the second closest.

SHIMURA: If it had swerved just a little from its orbit, it would have collided with the Earth.

KIGUCHI: Since it's a new comet, it's an important object of study.

SHIMURA: Is there any record of comets colliding with the Earth?

IKEDA: I remember reading somewhere that in the beginning of the twentieth century, a huge ball of fire fell on the Earth, landing in Siberia.

KIGUCHI: Yes, that was on 30 June 1908. It wiped out a forest area of about sixty kilometres.

IKEDA: Although the blast from the force was powerful enough to have derailed the trains on the Trans-Siberian railway, the mystery of the 'ball of fire' remained unsolved.

KIGUCHI: It wasn't until much later, on the basis of an analysis of fragments, that we came to believe that it was the result of a collision with the head of a small comet.

SHIMURA: Actually, the expression 'ball of fire' appears frequently in the records of the Kamakura period.

IKEDA: Aside from the *Azuma kagami*, the famous poet Fujiwara no Teika (1162–1241), who made a collection of favourite poems known as *Hyakunin isshu* (*Single Poems by a Hundred Poets*), recorded the event in his diary, *Meigetsuki* (*Journal of the Full Moon*).

KIGUCHI: Did Teika have any knowledge of astronomy?

IKEDA: He wasn't an astronomer, but he was living in the imperial capital of Kyoto and must have heard a lot about unusual heavenly phenomena being observed in many parts of the country. I think that his recording of some of these phenomena is a significant accomplishment.

KIGUCHI: What expression did Teika use for 'ball of fire'?

IKEDA: Here is the text. Would you read it for us, Mr. Shimura?

SHIMURA: 'At around the middle of the night, a shining object appeared in the heavens. Its size, force and intense brilliance suddenly appeared from nowhere, flinging itself from the southwest of the sky to the northeast, and then, in a matter of seconds, it burst open, spewing forth pieces of what seemed like burning charcoal.'

KIGUCHI: It has a different feeling from that of a scientist's description. In the field of archaeo-astronomy, the astronomical phenomena of those times are re-examined from the standpoint of the latest scientific findings.

IKEDA: It fascinates me that modern science is now verifying information recorded in the annals of history.

KIGUCHI: Among the specialists in this field of study, I'm sure there will be a few who will be most interested in probing the mysterious astronomical events which took place during this particular time period.

IKEDA: I once asked someone to look into the unusual heavenly phenomena which took place during the Bun'ei era (1264-1275). For example, one major event, according to the *Godai teiō monogatari* (*Tale of Five Imperial Reigns*), took place in the first year of Bun'ei, from the sixth to the ninth month. It was a comet which appeared and then disappeared a number of times, and its light spread halfway across the sky. The *Geki nikki* (*Diary of a Secretary*) makes note of a comet which appeared in the twelfth month of the second year of Bun'ei. The *Kantō hyōjōden* (*Chronicle of the Kamakura Shogunate*) records a comet in the first month of the third year of Bun'ei. The *Kintada-kō ki* (*Diary of Sanjō Kintada*) records another occurring in the seventh month of the fifth year. Moreover, during the span of the Bun'ei era, which totalled twelve years, there were a succession of solar and lunar eclipses, floods, droughts, typhoons, earthquakes, etc. The historical documentaion during this time period is filled with occurrences of such unusual phenomena.

KIGUCHI: These calamities, witnessed within such a short span of time, are very rare, even when viewed from the standpoint of astronomical occurrence on a worldwide scale.

SHIMURA: The eighth year of Bun'ei (1271) is worth noting because there was a remarkably high degree of unusual celestial occurrences.

Astronomical Phenomena and the Risshō ankoku ron

IKEDA: Nichiren Daishōnin referred to unusual occurrences in the heavens and Earth in his famous treatise, *Risshō ankoku ron* (*On Securing the Peace of the Land through the Propagation of True Buddhism*). In this work, he explains the reason for these occurrences as follows: 'When a nation becomes disordered, it is the spirits which first show signs of rampancy. Because these spirits become rampant, all the people of the nation become disordered.' 'Spirits' here mean evil or erroneous ideas that lead people down the path of disaster.

In more modern terms, the passage means to cause the positive functions of people's power of thinking to deteriorate, pulling humanity into one crisis after another. The people's lives become 'disordered' because evil thoughts cause people to deviate from the correct path toward happiness and prosperity. This confusion and the resulting social unrest and environmental pollution are all interrelated.

SHIMURA: It is essential that we recognize the relationship

between human society, the country as a whole and natural phenomena.

IKEDA: I am deeply interested in this particular writing which reflects the Daishōnin's profound understanding of this relationship.

KIGUCHI: It must be over ten years ago that I read something you wrote which stressed the importance of making this relationship the object of future academic pursuit and research.

IKEDA: I believe the chaotic age of the Bun'ei era in Japan, along with the European Middle Ages, signalled a significant turning point in the history of mankind. In its 'wake' are a multitude of problems which should be carefully considered and weighed by present-day writers and scientists in their avid quest for the ultimate truth underlying human existence.

The 'Shining Object' at Tatsunokuchi

SHIMURA: Looking back at history, Buddhas who expounded the Law almost invariably met with persecution. Was there any instance when a Buddha was killed?

IKEDA: Not in the case of Buddhas, but there are a number of instances when some of their contemporary and later disciples gave their lives for the sake of the Law.

KIGUCHI: It would seem, though, that Buddhas, who on occasion preached under very severe circumstances, would have been killed for certain.

IKEDA: That's understandable, particularly in the case of Nichiren Daishōnin who was almost killed at Tatsunokuchi. This incident is known as the Tatsunokuchi Persecution, and is documented in Japanese history books. But at the moment that he was to be beheaded, the executioner was unable to take the Daishōnin's life.

SHIMURA: Even during the Bun'ei era, with its rapid succession of unusual astronomical events, what took place at Tatsunokuchi is worthy of further discussion.

KIGUCHI: This particular incident is very famous, and although it happened some seven hundred years ago, the events as they were recorded appear to be accurate. In itself, the 'ball of fire' which shot across the sky over the execution site, lighting up all of Tatsunokuchi as if it were daylight, was astonishing enough. But it was an amazing coincidence that it occurred just when the Daishōnin was about to be executed.

SHIMURA: Based on the way it was documented, to what extent can we determine its validity? Modern people are inclined to analyze the reasons for this occurrence to determine whether or not there is actually a causal relationship.

IKEDA: As a follower of Nichiren Daishōnin's Buddhism, I interpret this event as being the manifestation of *hosshaku kenpon* (casting off one's transient identity and revealing one's true identity). In the case of the Daishōnin, at that moment in time, he discarded his transient status as a bodhisattva who was entrusted with the mission of propagating the Mystic Law by Shakyamuni Buddha, and revealed his true identity as the Buddha himself who is one with the ultimate Law of the universe. All the other Buddhas of the past, present and future from the ten directions of the universe derive their power from this ultimate Law which pervades the entire universe from time without beginning. This is the significance of the Tatsunokuchi Persecution.

SHIMURA: That's very profound, and yet it's very difficult to understand.

IKEDA: One can only understand this on the basis of his consistent faith and practice of Nichiren Daishōnin's Buddhism and one's continued study.

KIGUCHI: This point is actually the essence of our discussion on the relationship between Buddhism and the universe.

SHIMURA: What I learned about the Tatsunokuchi Persecution was that Nichiren Daishōnin, who had committed no offense, was sentenced to be beheaded by governmental authorities actually because he clarified and explained the reasons behind the frequent calamities and disasters which were taking place at that time by comparing the true and provisional teachings of Buddhism.

IKEDA: Yes, but there were also the false accusations lodged against him by priests of other sects of Buddhism. Traditionally, within the Buddhist community, a religious debate clarified the validity or invalidity of a particular sect. However, aware of the obvious superiority of the Daishōnin's teachings, the priests avoided debating with him or any of his disciples and instead plotted with high government officials to hinder his activities and even to take his life.

However, we can also view this from the standpoint of the *Lotus Sutra*, which predicted that in the Latter Day of the Law, persecutions would beset anyone who embraced and propagated its teachings. Nichiren Daishōnin appeared just at the beginning of this time-period and encountered all the persecutions predicted in the sutra.

KIGUCHI: In what part of the *Lotus Sutra* is the description of these persecutions to be found?

SHIMURA: In the *Kanji* (thirteenth) chapter, it is stated that one who propagates the *Lotus Sutra* in the Latter Day of the Law will be slandered and attacked with swords and staves and banished again and again.

IKEDA: Before the Daishōnin underwent the Tatsunokuchi Persecution, he had his home burnt down, an event known as the Matsubagayatsu Persecution (1260), was exiled in Izu (1261), was wounded on the forehead and had his left hand broken during the Komatsubara Persecution (1264), and, after his narrow escape from death at Tatsunokuchi, was exiled to Sado Island (1271), not to mention other kinds of persecutions.

Nichiren Daishōnin underwent all the persecutions predicted

in the thirteenth chapter of the *Lotus Sutra*. From one standpoint, this means that by fulfilling these predictions, he proved himself to be the Bodhisattva of the Earth entrusted with the mission of propagating the *Lotus Sutra* in the Latter Day, but at the Tatsunokuchi Persecution, the greatest of all the persecutions, he demonstrated his true identity as the Buddha of time without beginning, the Buddha who reveals and himself embodies the ultimate Law of *Nam-myōhō-renge-kyō*.

KIGUCHI: The Tatsunokuchi incident is very famous. When the executioner was about to lower his sword, a glowing object appeared in the sky, and, frightened, neither he nor anyone else was able to carry out the execution.

SHIMURA: As incredible as it seems, it's mentioned in the Daishōnin's writings given to his contemporary followers.

IKEDA: From the Middle Ages until now, numerous debates have taken place over the exact nature and actual existence of this 'ball of fire' which suddenly shot across the sky. Colouring these discussions, however, is, I feel, a lack of understanding of and a prejudice towards Buddhism, as well as a dearth of scientific technology needed to correctly ascertain the facts.

With regard to this episode, the details are described in such writings of the Daishōnin as the *Kaimoku shō* (*The Opening of the Eyes*) and *Shuju onfurumai gosho* (*On the Buddha's Behaviour*), and in his letters to Shijō Kingo, a devoted disciple, who, on hearing the news of the Daishōnin's impending execution, rushed to his side to die with him. The exact time was the eighth year of Bun'ei (1271), in the ninth month and thirteenth day,[4] between the hours of the Ox and the Tiger, which means between 2.00 and 4.00 am. According to these accounts, a luminous ball of fire flew from the direction of Enoshima Island which is located to the southeast of Tatsunokuchi and sailed across the pitch black of the night sky. Under the brilliant light, the executioner fell, was blinded, and the other samurai became panic-stricken and scattered in all directions.

A Japanese authority in the field of astronomy who

conducted research into this unusual phenomenon – the late Dr. Hideo Hirose (1901-1981) – used both historical and astronomical data to determine the facts, and his findings were that this glowing object was a meteor caused by the passing of Encke's Comet. He confirmed the date and time, and its elevation and angle of direction, which he extracted from historical documents.

SHIMURA: Dr. Hirose was a professor emeritus of Tokyo University and former director of the Tokyo Astronomical Observatory.

KIGUCHI: Encke's Comet has the shortest cycle of all the known comets, approaching the Earth at an interval of 3.3 years. American Astronomer Fred Whipple's study of this particular comet is well-known, and Dr. Hirose must have used it as the basis for his research.

IKEDA: The hours of the Ox and the Tiger, on the thirteenth day of the ninth month of the eighth year of Bun'ei in the Japanese lunar calendar translates into between 2.00 and 4.00 am on 24 October 1271, according to the modern calendar.

Dr. Hirose also used the German scholar, Karl D. Schoch's (1861-1913) table of the movement of heavenly bodies to confirm that the Moon had set at 3:44 am on that particular day. This corresponds to the passage in the *Shuju onfurumai gosho* which reads, ' "What if the dawn should break? You must hasten to execute me, for you will find it unbearable to do so after sunrise." I urged them on.' In *Ueno-dono gohenji* (*Reply to Lord Ueno*) he says, 'All the Buddhas of the past, present and future attain enlightenment between the hours of the Ox and the Tiger.' Dr. Hirose concluded that the 'luminous object' had to belong to the Aries-Taurus meteor swarm, which displays brilliant meteors at that time of year.

KIGUCHI: Encke's Comet is the source of the Aries-Taurus meteor swarm; Dr. Hirose arrived at his conclusion based on the calculation of the data concerning the location of the centre of the radiating meteor swarm caused by Encke's Comet, taking into consideration the differential with respect to the Earth's movement.

IKEDA: Based on Whipple's data, Dr. Hirose was able to confirm that this 'ball of fire' was a large meteor caused by Encke's Comet located in the Aries-Taurus constellation, occurring at 4:00 am, at an elevation of 34 degrees and positioned at an angle from south to west of 79 degrees.

KIGUCHI: Then Dr. Hirose's study does substantiate the Daishōnin's account.

SHIMURA: Incidentally, in May 1983, a scholar at the University of California at Santa Cruz in the United States confirmed the existence of a celestial body, probably the first planet to be observed outside our own solar system.

KIGUCHI: It was the first observation of its kind.

SHIMURA: In August 1983 there was also a noteworthy discovery of a solid body revolving around a fixed star outside our solar system, somewhere within the Milky Way.

KIGUCHI: Yes, around Vega, the principal star of Lyra. The American, British and Dutch tripartite-launched infra-red astronomical satellite (IRAS) detected a ring-like object made of large particles which is possibly a planet in its embryonic stage.

IKEDA: This is yet another new dawn in the space age, or rather, an age when we must truly awaken to the need to secure lasting peace on Earth.

SHIMURA: In order to alter the destiny of the world, which is presently dark and at a stalemate, toward one that is bright and filled with hope for the future, it is vital for us to have a revitalized view of our world and the universe around us, on the microcosmic level, which is our inner world, and on the macrocosmic level, which is our immediate environment and beyond.

Buddhist View of Illness

KIGUCHI: In connection with the subject of death, there is nowadays an increased concern for health.

SHIMURA: According to recent statistics (13 August 1983) published by the Japanese Ministry of Health and Welfare, one person in eight has some kind of serious illness.

IKEDA: Actually, illness has the most debilitating effects on life. I myself was preoccupied with the question of death since childhood, as I was almost always sickly.

KIGUCHI: In Buddhism the four sufferings of birth, old age, sickness and death are often discussed, but how is the relationship between sickness and death actually clarified?

IKEDA: While they are not perceived as being unrelated, they are not on an equal par either. Strictly speaking, these two phenomena are quite separate and distinct, and as Jōsei Toda used to say, 'Those who are good at being sick are usually poor at dying.' Medically speaking, serious illnesses like cirrhosis of the liver, cancer, cardiac arrest, stroke, etc, are likely causes of death.

KIGUCHI: How are these illnesses viewed from the standpoint of Buddhism?

IKEDA: Buddhism expounds that the body is composed of the four elements which are earth, water, fire and wind. Any disruption of these elements results in illness. In the *Lotus Sutra*, the Buddha is said to have 'few ailments and few troubles'. This means that since the Buddha's body is also composed of the four elements, and since he lives within a society which is filled with complex human relationships, he has the potential to become ill.

However, the Buddha has the power to recover the harmony of the four elements quickly, as well as to overcome spiritual suffering. This means he never succumbs to serious disease and suffering.

SHIMURA: What is meant by the 'four hundred and four sicknesses' mentioned in the Buddhist teachings?

IKEDA: It is a general term used to describe all physical illnesses caused by the disharmony of the four elements, which Nichiren Daishōnin mentions in his writing *Jibyō shō* (*Treatment of Illness*).

However, there are many diseases of the mind which are caused by the three poisons of greed, anger and stupidity, rather than by karma. They are said to amount to eighty-four thousand.

SHIMURA: What exactly do the four elements of earth, water, fire and wind represent?

IKEDA: They represent four types of energy flowing throughout the body. Earth indicates 'firmness' and has the function of holding the body together. In the human body, it indicates the bones, hair, nails, skin and muscles. Water is, of course, moist, and its energy serves to contain and connect all the parts of the body, and it comprises the blood and other bodily fluids. Fire provides heat and brings things to maturity, and thus regulates one's body temperature. Wind causes motion and thus acts to cause things to grow and to breathe. Illness is said to occur when these four elements become out of harmony with one another.

KIGUCHI: It's interesting to note that the first things a doctor checks on a patient are his temperature, blood pressure, pulse and breathing.

SHIMURA: Aristotle's theory of the four elements concerns only phenomena and is not as complete a system of thought as the Buddhist teaching which encompasses an overall view of human life itself.

IKEDA: It's also taught in Buddhism that sicknesses of the mind are more difficult to cure than sicknesses of the body. Buddhism assigns treatment of these illnesses to practice and faith in the Mystic Law.

SHIMURA: I understand that 'sickness of the mind' includes those caused by one's misleading thoughts and ideas.

IKEDA: That's correct. Mistaken views and evil thoughts themselves, which fall into the category of illusions or earthly desires in Buddhism, cause mental diseases. Nichiren Daishōnin clearly states that the most serious illnesses are caused by slander of the *Lotus Sutra*.

SHIMURA: We can say that maintaining the harmony of the four elements is, on a much larger scale, maintaining a state of harmony within the universe.

IKEDA: The universe is also comprised of the four elements or the four types of energy. When they fall out of harmony, aspects of the universe also become disrupted, and their restoration means that everything is again operating in rhythm. The Mystic Law not only brings about harmony within our own individual lives, but also within society, the world and ultimately the universe.

KIGUCHI: Among the stars which exist, there is a type referred to as a variable star, because it varies in brightness; it appears as if stunted in its growth, always 'gasping' for breath. The gravity and heat ratio is unstable, and the light emitted from it is irregular, so that, when viewed from the Earth, it appears to pulsate.

IKEDA: This seems to be an example of disharmony among the four elements. Buddhism holds that the four elements 'nourish' all things, so even a star in the vast universe is no exception to this rule.

How Life Is Regenerated

SHIMURA: Returning to the subject of death, rather than illness, can it be said for the most part that when one believes in the Mystic Law he will never commit suicide?

IKEDA: Not really. Nichikan Shōnin, said, 'Strong faith in the *Lotus Sutra* is in itself Buddhahood.' Nichiren Daishōnin stated, 'Without the lifeblood of faith, it would be useless to embrace the *Lotus Sutra* [the Mystic Law].' As these quotes clearly indicate, the strength of one's faith is the deciding factor in whether or not one will achieve enlightenment. Those people who practise with strong faith have every reason to want to live.

KIGUCHI: What about the accidental death of believers?

IKEDA: Of course this happens. Traffic accidents can be likened to what the Buddhist scriptures call 'being trampled by evil elephants'. Even if a person of strong faith dies as a result of an accident, this does not mean that he or she will fall into the three evil paths. In Buddhism, sending prayers to the deceased becomes especially important in such cases.

SHIMURA: After human beings die, how long does it take before they are reborn?

IKEDA: That's not an easy question to answer, and its resolution can only be arrived at within the dimension of faith. In Buddhism, it is taught that one who dies in the highest life-state will immediately be reborn as a human being, with renewed vitality and life force, and can contribute to society. In contrast, those people who have accumulated negative karma will fall into the three evil paths of Hell, Hunger and Animality, where they will remain for what they perceive to be an eternity.

According to some sutras, one's time in this 'journey' is said to be forty-nine days; every seven days, there is a chance to be reborn. This entity's time of rebirth is determined within the seven seven-day periods. This is the origin of the Japanese tradition of having services for the deceased on the seventh day after his or her death, and every seven days thereafter, until the forty-ninth day.

KIGUCHI: Enma (Skt Yama) is said to be the lord of Hell who judges the dead.

IKEDA: It's a figurative explanation of the world after death, which symbolizes conditions of happiness, fear and suffering that one's 'self' feels in the state of death. In any case, death is a reality, and there are many ways to die. Since everyone will inevitably face death, it is their strongest desire to die a peaceful death.

Nichiren Daishonin's Buddhism offers people hope and enables them to elevate their life-condition so that they can overcome all their sufferings. One can then face death without any regrets, even with a smile on his face.

I recall reading somewhere that no matter how many good deeds people may perform for society, it does not always mean that society will improve greatly. The author asserted that for society to really change for the better, another factor must be considered, and that is the salvation of those people who died tragic deaths. He also said that unless these people can experience enlightenment, an ideal society cannot be realized.

SHIMURA: In order to do this, reverence for the Law of life and the universe is necessary.

IKEDA: The subject, by its very nature, remains obscure, but I feel the sincerity of the author in his views on life and society. Apparently he must have realized that there is something far beyond the control of any human being. The things that the author has been discussing must remain within the limits of intellectual discourse, but, in Buddhism, one's prayer, that is, sending *daimoku* (*Nam-myōhō-renge-kyō*) for the repose of the deceased has deep significance and actual influence.

KIGUCHI: From a scientific point of view, it is extremely difficult to understand why offering prayers to the deceased is important, but I feel that there is something all-pervasive about doing this.

IKEDA: The daily practice of reciting the sutra and chanting *Nam-myōhō-renge-kyō* (called *gongyō*) enables one to make the blessings of the Law his own. At the same time, we are sending those blessings to the deceased every time we perform this prayer, morning and evening, as well as calling forth the positive forces, personified by all the Buddhist gods, such as the god of the sun, the god of the moon and the Four Heavenly Kings.

In his letter to his follower, Soya Kyōshin, Nichiren Daishōnin says, 'A beast called *mo*[5] feeds on iron. The earthly and heavenly gods, the dragon deities, the gods of the sun and moon, Taishaku, Bonten, the men of the two vehicles, bodhisattvas and Buddhas are sustained by the Buddhist Law, which becomes their flesh and spirit.'

SHIMURA: The expression, 'A beast called *mo* feeds on iron' reminds me that there are actually iron-eating bacteria which inhabit the hot springs and swamps.

KIGUCHI: There is hardly any living being who can consume iron, but these bacteria thrive by doing just that.

IKEDA: The Buddhist gods reveal their protective powers only after they have 'eaten' the blessings of the Law. In exactly the same manner, we can gain strong life force and courage, achieving harmony between mind and body, only when we continually 'nourish' ourselves with the blessings of the Law, the supreme Law of *Nam-myōhō-renge-kyō*, every morning and evening. I believe that doing so is the most meaningful and rewarding way to live.

NOTES: CHAPTER VI

1 Latent effect: the inherent effect in the depths of one's life, produced by an internal cause which is activated by an external cause. Since both internal cause and latent effect are dormant within life, they exist simultaneously, without the time differential that often occurs between an action and its manifest effect.

2 The Mongols attacked Japan twice: in October 1274 and May 1281. Prior

to the first invasion, Mongol delegates went to Japan, demanding that its government should acknowledge their authority. These representatives returned home safely, but the delegates sent during the time between the two invasions were killed, in 1275 and 1279.

3 *The Tale of Genji* translated by Edward G. Seidensticker, New York: Alfred A. Knopf, Inc., 1976.

4 Thirteenth day: it is commonly held that the Tatsunokuchi Persecution took place on the 12th. But strictly speaking, it was already the morning of the 13th when Nichiren Daishōnin was about to be executed.

5 *Mo* (Jap *Baku*): a legendary animal in Chinese culture. According to the classics, it is said to eat iron, copper and people's bad dreams.

VII

'The Threat of Existence' and the Mission of Buddhists

The forces of death and the power of life. Today's most pressing problem is the restoration of harmony with the rhythm of the universe.

The Spectacle of Death

SHIMURA: On 1 September 1983, a Korean Airlines jetliner was shot down by a Soviet fighter near Sakhalin Island and all 269 persons aboard were killed. The tragic downing of the airliner took on even greater significance and touched so many people because the passengers and crew who perished were nationals of fifteen countries. At that time, an editorial in the *Asahi Shimbun*, one of Japan's leading newspapers, stated, 'While we are aware that the times in which we live are dark indeed, that it should be this bad exceeds our imagination.'

IKEDA: Everyone throughout the world must have felt the same way. It was indeed heart-rending. I personally felt intense grief. Society may appear to be relatively peaceful, but in fact crises are breaking out everywhere. All people desire to live in peace with a feeling of security as they live their lives each day. Most of us go about our business from day to day never thinking that we may suddenly, unexpectedly meet death. However, in reality the forces of death are looming larger and larger, overtaking the forces of life. As described in the Buddhist teachings, this is characteristic of the age in which we live – the Latter Day of the Law.

KIGUCHI: The families of the passengers on board the ill-fated jetliner must have experienced a grief that is beyond the capacity of words to describe. One is again reminded that the question of life and death is an issue of the greatest importance.

IKEDA: Owing to the advances of medical science, today few people die from tuberculosis and pneumonia, diseases that were once considered incurable. Recent technological advances have enabled us to be forewarned about certain kinds of natural disasters. Consequently, the life span of human beings has increased somewhat; nevertheless, the people's life force has not. They are constantly faced with and being drawn toward the ever stronger pull of death. This is to be regretted.

SHIMURA: The number of people being killed as a result of air and traffic accidents is on the increase, and the incidence of such potentially fatal diseases as cancer, heart and kidney disease and diabetes is increasing. Aside from the inevitability of death itself, the level of people's anxiety is mounting because they are becoming more and more aware of the increasing difficulty in just being able to live a truly human existence.

My reaction to the KAL jetliner incident reminded me of the state of shock I had experienced when I heard that an *Asahi Shimbun* reporter, who had assisted me in gathering material for our magazine *Ushio* (*Current*), was killed in a plane crash on the outskirts of Moscow. I recall being in a complete daze that entire day.

KIGUCHI: Mr. Ikeda, have you ever witnessed accidental death at close hand?

IKEDA: Yes, a number of times in fact. When I was in primary school, our family was living in what is presently called Kōjiya in Ōta Ward in Tokyo. One day, on my way home from school, I saw a man lying motionless, face down in a pool of blood. Apparently he had been driving a truckload of construction materials, and he was wedged in between iron beams which had fallen from the truck. I did not know whether he was alive or dead, but the horrible sight was imprinted on my mind. Children are extremely impressionable and sensitive, and should not have to witness such things.

On another occasion, in 1948 or 1949, I saw a boy being run over by a bus and killed on a country road in Shizuoka Prefecture. His father turned pale as he rushed to the scene. He bent over his son's body and wept bitterly, calling his name and saying, 'Why did you die? Why did you die?' It is a memory that haunts me even now.

Some time before the Second World War, I lived near a river called Nomi-gawa. One particular Sunday, some people were fishing from a boat. One of them fell into the river, and almost immediately there was a great commotion. A crowd gathered at the scene. Eventually the body of the young man was recovered and brought up on the river bank. He appeared to be around twenty-five or twenty-six. I vividly recall that he was clad in a kimono.

Whenever I see an accidental death, I am reminded of the passage of the *Nirvana Sutra* which says, 'All is changeable, nothing is constant.' Nichiren Daishōnin says in the writing, *Shōgu mondō shō* (*Dialogue of the Sage and the Fool*): 'Neither young nor old know what fate awaits them – such is the way of our *sahā* world. All those who meet are fated to part again – such is the rule in this floating world we live in.' Impermanence is an inescapable fact of human existence.

SHIMURA: During the decade following the end of the Second World War, there were numerous tragic accidents in Japan, including the train fire accident at Sakuragi-chō Station in April 1951, the crash of the Japan Airlines plane *Mokusei* (*Jupiter*) in April 1952, and the sinking of the Aomori-Hakodate ferryboat, *Tōyamaru*, in September 1954. There seems to be no end to such tragedies.

KIGUCHI: Traffic accidents have become such routine occurrences that people have become insensitive to these small tragedies. This growing trend is appalling.

SHIMURA: In 1982, in Japan alone, the number of accidental deaths reached 9,073. The total number of accidental deaths worldwide must be staggering.

KIGUCHI: Moreover, if we add to this the people who suffer

from depression, alienation and lethargy, the numbers are astounding. In no other period in history has it been so difficult for people to lead normal lives. It would appear that the number of crises confronting humanity increases in direct proportion to the rate of scientific development. Indeed, many problems remain to be resolved, such as the constructive use of nuclear energy.

SHIMURA: War is the ultimate debacle where death reigns supreme. The forces of death are clearly present at such times.

IKEDA: During war people sometimes display great courage and compassion, more so than in peacetime – though not so much for the sake of preserving life as to squarely confront death.
 Courage should be the catalyst that brings forth the noble qualities inherent in human beings and which enables compassion to shed its light everywhere. In war, however, human valour degenerates into something brutal and insensitive, and is enlisted to engage in the act of murder based on misguided patriotism and the misconception that by such acts the defence of one's family and country is secured. This is truly the dark side of courage and humane intention.

SHIMURA: The positive and negative aspects of such virtues are determined by the way in which they are employed. Courage and compassion should be directed toward the realization of peace and happiness, and the protection of one's family. It is most unfortunate that these fundamentally positive qualities should be directed toward destructive ends.

The Desire to Eliminate Suffering

KIGUCHI: Another word for war is murder, but suicide, death by one's own hand, is also on the increase. No matter how you look at it, a kind of madness prevails in the world today.

SHIMURA: There are numerous instances in Japan of family

suicides where parents murder their children, prior to committing suicide because they feel that it is cruel to leave them behind. I suppose this is their idea of parental affection.

IKEDA: As Buddhists, we have the mission to completely eliminate such misery from the face of the earth. When filled with hope, one can motivate himself and even cause the surroundings to assume the tendency of an active, positive life force. On the other hand, when one is in the grips of despair, wallowing in the abyss of darkness, everything points toward death. Therefore, in determining the course of one's life, it is most important to firmly establish one's *ichinen*, or single-minded devotion, to life. This *ichinen*, though not externally discernible, becomes a driving momentum within one's life. One can achieve this only through faith in the absolute Law of the universe.

SHIMURA: I have become aware that the strength of one's will to live is of prime importance. But lately the growing tendency has been to choose death instead. What causes people to feel such unbearable pain and sorrow that they choose death as the only alternative?

KIGUCHI: Pain and suffering no doubt are causes for the decision to commit suicide, but more to the point, I believe that they have lost all hope. This brings to mind the proverb, 'Where there is life, there is hope.'

IKEDA: My mentor, Jōsei Toda, once said, 'During the course of the continuous cycle of birth and death, a person's life may become defiled because of bad habits or life-styles based upon their misconceptions. Such habits come from earthly desires that take the form of the three poisons of greed, anger and stupidity. Once one's life becomes distorted, he will be out of rhythm with the universe, and then his life force will become weaker. Consequently, the person becomes less and less able to respond positively to the various situations which arise in his surroundings, and, as a matter of course, living itself becomes very painful and tragedy results.'

SHIMURA: Would you explain 'greed, anger and stupidity' in more detail?

IKEDA: Greed characterizes a selfish ego and is a state of mind that lacks any sense of gratitude. Therefore, such people are always consumed with indignation, regardless of the favours or kindnesses that they receive from others. They are never satisfied. While some people who are controlled by insatiable desire may enjoy fame and honour, in the light of Buddhism, they are in the state of Hunger, and are dominated by selfishness and greed.

KIGUCHI: Buddhism is indeed strict. Moreover, today people are developing the tendency, unfortunately, to become angry at even the most trivial matters.

SHIMURA: I understand that the Chinese character for anger originally meant to open one's eyes widely when glaring at another.

IKEDA: That's correct. This life-condition is very close to the world of Hell in which people are controlled by intense anger. People in this life-condition become angry at anything and always put the blame on others or on their surroundings, even when, in fact, they themselves are to blame. Their lives are consumed and their hearts overtaken by anger, and they then fall into the state of Hell.

SHIMURA: In Buddhism, 'stupidity', one of the earthly desires, means complete ignorance of all things that are logical.

IKEDA: Nichiren Daishōnin equates stupidity, a state dominated by instinctive desires, with the world of Animality, saying, 'The wise may be called human, but the thoughtless are no more than animals.'

KIGUCHI: Through the understanding revealed by Buddhism, it is possible to comprehend the life-condition that results in the desire to commit suicide. The fact that the number of suicides is increasing indicates that people today are finding it more and more difficult to live, having little freedom to move around in, and that their life force is diminishing.

SHIMURA: In Buddhism, the conditions of greed, anger and stupidity are called the 'three poisons'. Why is this so?

IKEDA: The *Lotus Sutra* states, 'The poison has penetrated deeply,' [causing them to lose their true minds].' Poison, by its very nature, is destructive and causes the deterioration of the human being's innate goodness. Just as the body is affected by the ravages of physical illness, the mind is subject to the ill effects of the three poisons. The *Daijōgi shō* (*Buddhist Concepts Viewed in Light of the Mahayana*) says in effect that these poisons give rise to all illusions and earthly desires in the threefold world, where unenlightened beings live. These poisons, extremely virulent, wreak havoc on one's mind. It is said that, like pathogenics such as bacteria, the poisons work not only to eat away at one's existence in the present lifetime, but their polluting effects continue into future lifetimes unless the person embraces the Mystic Law.

KIGUCHI: It follows, therefore, that even if one commits suicide, he cannot escape from suffering because death is not final.

IKEDA: That's correct. Shakyamuni himself said that the Latter Day of the Law would be an evil age, ridden with the five impurities. Just as he predicted, modern society has degenerated to the extent that the people's lives are polluted by the three poisons.

KIGUCHI: What are the five impurities?

IKEDA: They are outlined in the *Hōben* (second) chapter of the *Lotus Sutra*. To explain briefly, they are the impurities of the age, of society, of earthly desires, of thought and of life itself. The Daishōnin elucidates this further in the *Jibyō shō* (*The Treatment of Illness*) as follows: 'The heart of the Hokke

[Lotus] sect is the principle of *ichinen sanzen* [a single life-moment possesses three thousand realms], which reveals that both good and evil are inherent even in those at the highest stage, that of *myōgaku* or enlightenment. The fundamental nature of enlightenment manifests itself as Bonten and Taishaku, whereas the fundamental darkness manifests itself as the Devil of the Sixth Heaven.' Bonten and Taishaku are two main tutelary gods, and the Devil of the Sixth Heaven is the king of devils (devilish forces) who delights in sapping the life force of human beings.

In other words, as the Daishōnin teaches in the *Tōtaigi shō* (*The Entity of the Mystic Law*), the existence of both good and evil, or enlightenment and delusion inherent within human life, notwithstanding, the life-essence underlying both is one and the same. When one's life-essence encounters evil influences, it manifests in the form of delusion, and if it encounters positive influences, one's enlightenment is activated.

Today, in the Latter Day of the Law, the people's minds are tainted by the three poisons, the fundamental negative forces inherent in life; as a result, their lives are completely defiled.

SHIMURA: Can it be said then that the impurity of life is the most basic among the five impurities?

IKEDA: Yes, but we should also be aware of the interrelationship among the five impurities themselves. In other words, it is necessary to understand the impact of each impurity on the other four, based on the principle of *eshō funi*, or the oneness of life and its environment. Shakyamuni Buddha, the Great Teacher T'ien-t'ai and Nichiren Daishōnin all taught that the fundamental darkness inherent in human life is the root of all unhappiness, and that to penetrate this darkness is to bring forth life's innate goodness, resulting in one's personal happiness and the creation of a peaceful society.

KIGUCHI: It has dawned on me that while astronomers are preoccupied with investigating the vast expanse of the universe, light is beginning to shine into the innermost depths of human life. In other words, deep within each individual's life is a microcosm which Buddhism terms *ichinen*.

IKEDA: Controlling the evil side of our nature and drawing forth the goodness from the depths of our lives is the noblest way to live as human beings. We should strive to do so throughout eternity, regardless of how prosperous or advanced our surroundings may be. There is no other way to stay on course and to cultivate and preserve human happiness. For our own sake we must adopt the correct view of life and death. Moreover, we must do so if we wish to create lasting peace in the world. If we become too preoccupied with daily concerns, time will pass by all too quickly, and we will have achieved nothing truly noteworthy. As the saying goes, 'Time flies like an arrow.'

KIGUCHI: Life span is defined as the period between birth and death. The stars and the planets throughout the universe also have determined life spans. How might we regard the life span of a human being in terms of Buddhism?

IKEDA: In general, Buddhism teaches that various forms of life make their appearance in this world, sometimes defined as the threefold world in which unenlightened beings transmigrate within the six paths, and that these forms of life have life spans which are fixed or determined. In Japanese, a fixed life span is translated as *juryō*.

SHIMURA: What, then, does *juryō*, in the *Juryō* chapter of the *Lotus Sutra*, mean?

IKEDA: Briefly, it means eternal life, or life which continues throughout the three existences – from time without beginning through the present and into the infinite future. Although *juryō* can indicate the life span of a human being, in this chapter of the *Lotus Sutra* it refers to the life span of the Buddha who exists throughout eternity.

There are cases where an individual's life span is predetermined, based on his or her karma. This is called *jōgō* or immutable karma. However, one's life span may be shortened even further because of illness or other factors caused by negative influences. On the other hand, Buddhism teaches that the power of the Mystic Law enables one to change his negative karma and thus prolong his own life.

SHIMURA: Speaking of life span in terms of a number of years, would it be possible to measure the length of a person's life span by taking his karma into account? It has been said that the physical body can be likened to a clock winding down. When the clock stops, we die. Perhaps a more apt analogy would be that of a battery running down.

KIGUCHI: There are biologists who are presently attempting to find out whether or not an individual's life span is in fact predetermined.

SHIMURA: Yes, for example, a West German scientist has conducted experiments in which the heartbeats of elephants, one of the largest animals in the world, and mice, one of the smallest, were monitored and compared. According to his report, the average number of heartbeats for elephants during their lifetimes is 1.02 billion, while that for mice is 1.11 billion. Elephants have an average life span of seventy years and mice have approximately three and a half years. Although the life span of the elephant is twenty times that of the mouse, the number of heartbeats of these two animals is very nearly the same.

KIGUCHI: This may well mean that the life-energy latent in animals can somehow be determined.

The Traditional Japanese View of Life and Death

SHIMURA: In 1983, the announcement of those people who had lived to be a hundred years of age or older was given official recognition. Longevity is something everyone would like to celebrate.

IKEDA: People who have lived for such a long time are to be admired and congratulated. I've heard that in Okinawa, it has long been a tradition for women and children to visit the homes of such people on the morning of New Year's Day.

SHIMURA: The Japanese Imperial Household makes offerings for those deceased persons who during their lifetime made significant contributions to society. Red and white rice cakes are given out, particularly at the funeral services of those who had lived to a very old age. The Imperial Household, which has played a central role in the preservation of Japanese traditions since ancient times, has assumed the responsibility of maintaining the custom of offering congratulations to people reaching old age.

IKEDA: Everyone would like to live to a ripe, old age and die a natural death. How one can achieve this is indeed an intriguing question.

SHIMURA: It is interesting that the Japanese language has many different ways of referring to 'death'. For example, the word *shikyo* means to pass away, *shibō* means to be deceased, and *eimin*, eternal sleep.

IKEDA: According to classical literature, the word which applies to the death of the Buddha, *nirvana*, differs from the one which applies to the death of an emperor, a feudal lord, a man of learning or a shogun. There are documents which refer to the deaths of commoners as *shi*, meaning death, and sometimes *enkō*, which means 'going to a faraway place'. *Hei*, to collapse, refers to the deaths of animals; this word was also used for those who have wronged others and are therefore considered to be on the same level as cattle. One hears the saying, 'If one commits many wrongdoings, he is certain to collapse and die.'

SHIMURA: Why does the Japanese language have many words and expressions referring to death?

IKEDA: That is an interesting question; I would imagine that while the Japanese language has been strongly influenced by Chinese culture and thought, the usage of many terms to express this one concept reflects the characteristic sensitivity with which Japanese people view the phenomena of life and death.

SHIMURA: Apparently, for there are few countries in the world where memorial services are held even for insentient objects like needles as they are in Japan.[1]

IKEDA: The sensibilities of the Japanese people concerning death are subtle and delicate. For this reason, the use of but a single word to express this, the ultimate problem for human beings, would not be acceptable. One Chinese character for death consists of two radicals: the left radical is *gatsu*, an ideograph depicting bones crumbling, and the right one is *ka*, meaning 'to drop dead'. Together, they signify life reaching its end.

SHIMURA: Moreover, there are words like *kon*, or soul, and *haku*, or spirit, both of which indicate forms of life after death. The radical on the right side of each of these characters is called *ki*, or the spirit of a dead person.

IKEDA: The character for *ki*, which also means demon, is said to have originally indicated the form of a deceased person. It would appear to symbolize the anguish of dying. Because many people have died in agony, perhaps this character may be understood as expressing the look of unbearable suffering which has appeared on the faces of people at the time of their death.

SHIMURA: Just the sight of a person dying saddens that person's relatives and friends. Perhaps this is because of the interaction between living beings and their surroundings.

IKEDA: Moreover, when a person dies a miserable death, it is painful for us to witness the person's family experiencing deep sorrow and grief. It is no wonder then that since antiquity, death has been looked upon as the thing to fear the most.

SHIMURA: Many types of rituals originated in the fear of death.

IKEDA: One belief is that the tortured spirit of a person who died in agony can cast evil spells and wreak havoc on the living.

KIGUCHI: I've heard that in some localities, when a person dies young or meets with accidental death, it is the custom to worship that person as a deity of youth.

SHIMURA: Then there is the worship of spirits, a practice which is thought to bring repose to the souls of those who died unnatural deaths or died while bearing grudges. A well-known example is the worship of the spirit of Sugawara no Michizane, a leading scholar and court official of the Heian period (794-1185), who challenged the powerful Fujiwara family, but was betrayed and died in disgrace. Another is Taira no Masakado, a warrior of the Heian period who led the first major rebellion against the government, and Sakura Sōgo, a farmer and folk hero during the early Edo period (1600-1867). Shrines were built and dedicated to them after they died.

KIGUCHI: The Nembutsu sect, which expounds rebirth in the Pure Land of Amida Buddha, is said to have acquired a large following by exploiting the common people's ignorance and fear of death.

IKEDA: According to the teachings of this sect, this world abounds with impurities and sufferings. It is an unfortunate historical fact that many people committed suicide, believing that by reciting the name of Amida Buddha, they would be able to escape this world and be reborn in the Pure Land of Perfect Bliss. Such acts were nothing less than insane.

KIGUCHI: Even today, many sophisticated and educated people avoid doing anything important on so-called unlucky days, such as *butsumetsu*, the day of the Buddha's death, or *tomobiki* (literally meaning to draw a friend to the land of death), a 'trail' day which deviates from the Yin-yang calendar, or they rely on astrology. This must mean that most people have not as yet been able to transcend their ignorance and fear of death.

IKEDA: Such customs perhaps derive from ancient Chinese teachings, and are regarded by many people as only folk beliefs or superstitions.

SHIMURA: It is said that the Chinese character *butsu* from *butsumetsu* originally signified 'things', but eventually it was replaced by the homonym meaning 'Buddha', partly because the latter character had a more august meaning and therefore found acceptance among the common people. Based on this reasoning, some scholars dismiss the validity of the common interpretation of *butsumetsu* as being groundless. To give an extreme example, in a certain locality in Okayama Prefecture, people still believe that when an elder brother dies, his spirit will beckon his younger brother to the land of the dead; in order to prevent this, the younger brother changes his name.

IKEDA: Nothing has distorted people's minds to the point of absurdity more than the fear of death. Of course, this is not to say that these ways of thinking are erroneous because they're old. In fact, many ideas and thoughts which originated in the past continue to create value and generate wisdom. However, it is also the case that, in spite of the fact that civilization has made considerable progress, and the fields of politics and economics have brought about great changes, the ancient views of death are persistent and remain deeply rooted in the depths of human consciousness.

SHIMURA: Folklorists say that the influence of ancient cosmological concepts still prevail, such as the notion of an eternal or nether world.

KIGUCHI: Cosmological concepts were an integral part of ancient astronomy. Where those views have survived until today, they are usually considered mere legends and folklore, but when it comes to views concerning death, the inevitable destiny of every living being, people's minds remain bound up in superstition.

IKEDA: Modern man tends to avoid confronting this issue, but the question of life and death is one that cannot be put off indefinitely. It is imperative that we find a solution to this problem.

KIGUCHI: Regardless of one's position in society, as a human being, one is destined to die. Only the clarification of the true nature of death can enable people to experience true happiness and to live their daily lives confidently.

IKEDA: Naturally, the inevitability of death makes people want to escape it. The fear of possible anguish or agony before death is dormant in most people's minds.

SHIMURA: Every year there is an official holiday set aside to acknowledge respect for the aged in Japan; on television, Prime Minister Yasuhiro Nakasone discussed matters concerning old age with several people including Yūji Aida, professor emeritus of Kyoto University. The main topic was how to make one's old age as comfortable as possible, but the question of death was not even touched on. In fact, not everyone's death is traumatic. There are many people who die in a state of tranquillity and peace, after having lived out their lives with no regrets.

IKEDA: It would be comforting to believe that we will die calmly, as if we were falling asleep. Nevertheless, many people accept death with a fatalistic attitude and regard it as something to be feared, having heard about or witnessed deaths that were traumatic and painful. But we should keep in mind that many people face death peacefully. It is my belief that faith in the Mystic Law is what enables people to feel reassured. Those who have a positive view of death and lead invigorated lives will certainly have the kind of fulfilling lives that escape those who yield to external influences and are preoccupied with cunning and selfish pursuits.

No one can escape death. A person may meet death at any time; this is the fate of every single human being. Nevertheless, this fact is very difficult for people to really accept, probably because they are overwhelmed by the daily pressures of work and other responsibilities. But those who do wish to understand the essence of life must deal with the issue of death; it is a necessary step in the process of realizing life's true meaning.

SHIMURA: A letter from an elderly man was published in the *Asahi Shimbun* several days ago. He wrote to the effect that many people visit him at the old-age home where he resides and try to comfort him, but, as he says, 'None of them can tell me anything about death, something which we all fear.'

IKEDA: His words are very revealing.

KIGUCHI: There are certainly enough people today who talk about death, but no one, I'm afraid, can offer us any insightful conclusions.

SHIMURA: There are shrewd people who try to profit from people's fear of death. A number of religionists even try to commercialize death, but they themselves cannot resolve the life-and-death issue.

The Mysteries of Venus

SHIMURA: I am told that the position of Venus will be at its closest to the Earth on 1 October 1983, and the planet will be clearly visible even in the daytime.

KIGUCHI: In relation to the Earth, Venus is the third brightest heavenly body, next to the Sun and the Moon. It is exposed to twice as much sunlight as the Earth, thus reflecting that much more light. It is also the closest planet to the Earth.

IKEDA: In the spring of 1982, a Russian spacecraft soft-landed on Venus's surface and succeeded for the first time in taking colour photographs of the planet, in temperatures high enough to melt lead. What about its brightness during the daytime?

KIGUCHI: We aren't completely certain yet, but the

photographs indicate that Venusian daylight is about as bright as an early afternoon on a cloudy winter day in Moscow.

IKEDA: Why is it, then, that there is not so much light at the planet's surface, when the planet receives twice as much sunlight as the Earth?

KIGUCHI: Venus is shrouded in black layers of atmospheric clouds composed mainly of carbon dioxide and water vapour as the Earth was in the early stages of its evolution. The gas has permeated most of the rocks on the planet's surface, and formed compounds such as limestone.

When the Earth's surface cooled, the water vapour in the atmosphere condensed into liquid and formed oceans. At the same time, the carbon dioxide in the atmosphere combined with the water. The outcome of this was the formation of rock on Earth as we know it.

SHIMURA: Why is there so much carbon dioxide present in Venus's atmosphere?

KIGUCHI: The heat generated by the intense sunlight causes the water vapour to dissociate into hydrogen and oxygen gases. Hydrogen being a light element, it escapes into space, but the oxygen and carbon dioxide remain in the atmosphere. However, the oxygen eventually settles down on the surface of the planet and oxidizes it, resulting in the planet's surface having a reddish-brown colour. Consequently, only carbon dioxide is abundant in the atmosphere.

IKEDA: The heat emanating from the Sun determines whether or not water is formed, and the presence or lack of water determines a planet's destiny. Is that correct?

KIGUCHI: Yes, and since Mars is farther away from the Sun than the Earth is, the water on Mars is mostly frozen within its surface.

IKEDA: I understand that Venus exhibits different phases, like the Moon.

KIGUCHI: That's right. When Galileo observed this phenomenon through his telescope, it further strengthened his conviction about the heliocentric theory, despite the overwhelming opposition of the Church to such a theory at the time.

SHIMURA: I have read that when Venus moves closer to the Earth, the same hemisphere always faces the Earth, like the Moon. Why is this?

KIGUCHI: The same hemisphere of the Moon always faces the Earth because the energy of the Moon's rotation on its axis becomes diffuse and dissipates, due to the gravitational pull of the Earth. It's interesting to note that although Venus is heavily influenced by the pull of the Sun, it is also thought to be affected by the Earth as well. However, there is no concrete proof of this at this time.

SHIMURA: It is said that one year on Venus is about two Venusian 'days long'.

KIGUCHI: If we define the length of one day as the time between when the Sun rises and it rises again, then this is correct. This is because the length of time that it takes for Venus to rotate (retrograde) is roughly equal to the time it takes for it to revolve around the Sun, which is about two hundred and thirty-four Earth days.

IKEDA: In other words, Venus rotates much more slowly on its axis than the Earth does, but revolves more quickly around the Sun, because its path is shorter. Measurement is adjusted according to the speed of a given planet's rotation which determines the length of its day, and the speed of its revolution around the sun, which determines the length of its year.

SHIMURA: According to a treatise written by Fred Hoyle, a British astronomer, the Earth, in its infancy, rotated on its axis once every five hours. In other words, one day was 5/24 as long as our present day.

KIGUCHI: This means that the Earth's rotation has slowed down about five hundred per cent.

SHIMURA: It is interesting to note that the present length of time is optimal for sustaining of life.

KIGUCHI: One day on Saturn is ten hours long.

IKEDA: The perception of time and space is not absolute. It differs according to what is being observed and its position relative to the observer. The results of Dr. Hoyle's calculation of the distance between the Earth and the Moon fascinate me. It was observed that the Moon was once one third closer to the Earth than it is now, and is gradually pulling farther away.

KIGUCHI: The Moon is still retreating at the rate of about three centimetres a year. Some astronomers say that there is the likelihood that the Moon will break away completely from the Earth in the far distant future.

The Falling Star at Echi Was Venus

SHIMURA: Venus is referred to in Buddhism as the god of the stars.

IKEDA: The planet is a symbol of the benevolent workings of heavenly deities. Since ancient times, it has been revered as the 'morning star' or 'evening star' in both Japan and abroad.

KIGUCHI: Venus is so clearly visible that it is sometimes also called the 'flying bright star' or 'jumping bright star'.

IKEDA: An old fisherman living at Katsuura in Chiba Prefecture was reported to have said of the morning star: 'It "jumps up" about eighteen feet above the horizon at about three o'clock in the morning.'

SHIMURA: An old man living in Hachiōji, Tokyo, a town once

famous for its manufacture of silk goods, said that as a child he and his friends would view the planet through a piece of silk cloth because the outline of the planet could be seen more clearly.

KIGUCHI: That's because silk cloth or bird feathers exert a diffraction effect on light emitted from the stars.

IKEDA: The *Nippon tenmon shiryō* (*Japanese Historical References on Astronomy*), an impressive work edited by Shigeru Kanda, states that in the *Shoku nihongi* (*Continuation of Chronicles of Japan*) alone, there are as many as ten passages which mention stars appearing in the daytime.

SHIMURA: Venus emits a very bright light. I am reminded of an unusual astronomical phenomenon that occurred during the time of Nichiren Daishōnin. The day after the Daishōnin was almost beheaded at Tatsunokuchi, he was held in custody at the Honma residence in Echi near Atsugi in present-day Kanagawa Prefecture. In the garden, he recited the sutra and admonished the deities. Immediately afterwards, a large bright star 'fell from the sky' and appeared to be suspended in the top branch of a nearby tree.

IKEDA: The Daishōnin described this falling star in the *Shuju onfurumai gosho* (*On the Buddha's Behaviour*): 'Because it was the middle of the ninth month, the Moon was very round and full.' It was probably the Moon in the thirteenth day of its cycle. He goes on to say, 'I went out into the night garden and there, turning toward the Moon, recited the *Jigage* portion of the *Juryō* chapter. Then I spoke briefly about the merits and faults of the various sects and about the teachings of the *Lotus Sutra*.' In other words, the Daishōnin was reading the sutra facing the Moon.

Nichiren Daishōnin says in the *Ongi kuden* that the verses of the *Jigage* reveal the truth that the life of the Buddha is eternal without beginning. It clarifies that the enlightened life of the Buddha continues to exist from the infinite past to the eternal future. The fundamental principle or ultimate Law underlying this sutra is the Law of *Nam-myōhō-renge-kyō*. In his writing, the *Hōren shō* (*Letter to Lay Priest Hōren*), Nichiren Daishōnin teaches, 'The eyes of ordinary mortals see the characters of the *Lotus Sutra* only as black characters, while the men of the two vehicles regard them as *kū* (non-substantiality) and bodhisattvas look on them as innumerable teachings, but the people in whom the seed of Buddhahood is ripe recognize each character of the sutra as a Buddha.'

SHIMURA: Returning to the scene at the garden of the Honma residence, after reciting the sutra, the Daishōnin strongly admonished the heavenly gods who represent the protective forces and functions of the universe, saying, 'Now that you see me in this situation, you should joyfully rush forward to shield the votary of the *Lotus Sutra* and thereby fulfil your vow to the Buddha ... What is your answer, Moon? What is your answer?'

IKEDA: The *Anrakugyō* (fourteenth) chapter of the *Lotus Sutra* states, 'Day and night, for the sake of the Law, the heavenly gods will constantly guard and protect him [the votary of the *Lotus Sutra*].' In addition, the *Zokurui* (twenty-second) chapter also says, 'Transmit and protect the *Lotus Sutra* as the Buddha commands.'

SHIMURA: The Daishōnin states that as soon as he finished admonishing the heavenly gods, a large bright star fell from the night sky and 'struck' a plum tree in the garden.

IKEDA: He also wrote in a letter dated the twenty-first day of the ninth month in the eighth year of Bun'ei, to Shijō Kingo, a samurai and devoted follower of the Daishōnin, about this strange phenomenon, 'The god of the stars descended four or five days ago to greet me.' With regard to this particular falling star, Dr. Hideo Hirose, former director of the Tokyo Astronomical Observatory, wrote an intriguing report. He said that when he read the above-quoted passage, he immediately thought that it might possibly be none other than the planet Venus. He began to make calculations in order to confirm his suspicions. Whatever the case, we should bear in mind that the occurrence of this phenomenon at that particular moment has a significance that goes beyond the scope of astronomy and enters the dimension of Buddhism. Otherwise, it is just a mere historical fact.

This particular phenomenon occurred on the thirteenth day of the ninth month in the eighth year of Bun'ei, which, corresponds to 24 October 1271 in the Western calendar. Dr. Hirose looked up the position of Venus and determined its course in the fundamental catalogue, devised by the German astronomer Karl D. Schoch. The results of Dr. Hirose's findings, based on the data he accumulated for Venus, revealed that, on that day, the planet would have attained its maximum brilliance (magnitude -4). In other words, it must have been about a hundred times brighter than a star of the first magnitude.

So, on the day the star 'fell' in the western sky, the light reflected by Venus was at its brightest. Dr. Hirose calculated that on that day it was in the evening star phase and the Sun had set around five o'clock in the evening; Venus could have been seen for only two and a half hours after the sunset.

Dr. Hirose reconstructed the events of that night as having happened this way: soon after the sunset, the eastern sky became clear, and the Moon, in the thirteenth day of its cycle, appeared. In the western sky, the clouds hung low, so Venus could not be seen. As soon as the Daishōnin completed the recitation of the *Jigage* and then declared his admonishment, the planet Venus suddenly appeared from behind the clouds and cast its brilliant light above the plum tree. But a passage states that the sky was soon darkened by clouds. They must have gathered in the western skies. Dr. Hirose concluded that the event must have taken place within a very short period of time.

I realize that I am not qualified to discuss this astronomical phenomenon on the basis of only one source. I am merely speculating, but this explanation seems to scientifically support the Daishōnin's account.

KIGUCHI: I believe that Dr. Hirose's findings are indeed impressive, and, notwithstanding the criticisms which may be lodged against them, are nevertheless worth considering.

SHIMURA: With respect to the falling star referred to in the Daishōnin's writings, I'm told that a 'twilight arc' is sometimes seen when Venus is observed.

KIGUCHI: That's right. From time to time, a ring of diffused light can be observed around the entire planet.

IKEDA: It is just as the old fisherman said: Venus sometimes suddenly appears and seems to sparkle. It is also conceivable that at its maximum brilliance, Venus could shine so brightly as to cast a shadow in the same way that the Moon does.

KIGUCHI: It is then conceivable that, on seeing such a phenomenon, one might think he has in fact seen a falling star.

NOTES: CHAPTER VII

1 On 8 February or 8 December, or both days in some parts of Japan, women collect all their old needles. Sticking them into soft cakes such as *tōfu* (bean curd), they pray for the repose of the needles, improvement in their sewing skills, and safety from any injuries while sewing.

VIII

Life and Death: the Last Frontier

Human spiritual revival must be based on the Buddhist view of eternal life, which teaches that both 'life' and 'death' are the functions of life itself.

One's State of Mind Colours the External World

SHIMURA: The setting sun sometimes appears larger and redder than it does at other times.

KIGUCHI: That phenomenon is the result of an optical illusion. In actuality, the Sun is the same size regardless of its position in the sky.

SHIMURA: Could it perhaps be changes in the atmosphere which make the Sun look bigger?

IKEDA: I occasionally take photographs of the setting sun and the Moon, and sometimes they appear incredibly large and beautiful.

SHIMURA: I know of a scientist who is conducting research concerned with the relationship between human beings and heavenly bodies. Among other things, he is taking a survey of people's conception of how large the Sun appears to be, in centimetres.

KIGUCHI: That's quite interesting. Astronomers have not concerned themselves with such matters, but I believe that this type of inquiry will prove to be an important area of study in the future.

SHIMURA: He found that when viewed, the Sun appears to be about thirty to forty centimetres in width to most people, but appears larger to those individuals who are broad-minded or are involved in creative occupations. On the other hand, it looks smaller to those people who tend to be argumentative or whose jobs require precise calculations.

KIGUCHI: Thus, a person's state of mind has an influence on how he or she perceives the Sun.

SHIMURA: One respondent in this survey said that the Sun appeared to be about a metre in diameter.

IKEDA: The different perceptions are clearly a reflection of each individual's life-condition.

The Influence of the Moon on Human Behaviour

SHIMURA: The book *Supernature* by Lyall Watson reports an interesting study on the influence of the full moon on human behaviour issued by the American Institute of Medical Climatology. It reports that the number of crimes of psychotic origin reaches a peak when the Moon is full.

KIGUCHI: Two hundred years ago, British criminal law made a legal distinction between crimes which were committed by insane individuals who had incurable chronic mental disorders, and those committed by people who purportedly experienced temporary mental aberrations because of the Moon's influence.

IKEDA: Does this mean that those crimes which were committed during the full moon were dealt with more leniently?

KIGUCHI: It would seem so. And mental institutions during that time period prohibited their staff from leaving the asylums on nights when the Moon was full. It is also said that, during the eighteenth century, patients would be beaten the day before the full moon rose as a preventive measure against their going on a rampage during the night of the full moon.

SHIMURA: Does the Moon actually have any influence on human beings?

KIGUCHI: The waxing and waning of the Moon is caused by the relative movements of the Earth, Moon and Sun. These positional changes affect the pull of gravity as well as its direction. Perhaps this gravitational shift has an impact on the blood flow and electrical potential in the human being.

There is the belief that the Moon may influence blood circulation in the same way that it affects the tides of the ocean. According to one medical report, bleeding crises, such as haemorrhaging, reach a peak during the full moon phase. This seems to suggest a relationship between the Moon and human beings.

SHIMURA: The study concerning the relationship between human beings and the Moon is still in its early stages. However, there are many phenomena which seem to point in that direction, compelling many to conclude that indeed there must be some kind of connection. In some places, the Moon is said to be called 'the great midwife' because of its apparent influence on childbirth.

There is also a medical report that deaths from tuberculosis are most frequent seven days before the full moon's advent. In this report, it was conjectured that the phases of the Moon might exert an influence on the ratio of acid to alkali in the blood.

Men of Literature Who Confronted Death

SHIMURA: A certain ancient Greek philosopher said that only when one is prepared to meet death is he free in the truest sense of the word. This may be idealistic, but we must bear in mind that every individual is destined to die someday.

IKEDA: Since no living person can escape death, he should make every effort to come to terms with its reality.

SHIMURA: An episode in the life of the Japanese writer, Tōson Shimazaki (1872-1943), is worth noting.

Tōson visited a writer, Katai Tayama (1871-1930), who was on his deathbed. At this crucial time, Tōson asked him, 'How does it feel to be at the point of death?' The family members who were there at the time all frowned at him. Later on, Tōson apologized, explaining that he wanted to understand a fellow writer's true feelings and thoughts about death since both of them had pursued the world of literature, and there was simply no one else to ask.

KIGUCHI: As a literary man, Tōson must have wanted to gain some insight into the reality of death.

IKEDA: Indeed, many people of literature have tried to probe the question of death, for example, Proust, Dostoyevsky and Hemingway, to name a few.

SHIMURA: Death has been an issue in the works of such Japanese writers as Sōseki Natsume (1867-1916), Ryūnosuke Akutagawa (1892-1927), Naoya Shiga (1883-1971) and so on.

IKEDA: There is a famous *tanka*, a Japanese verse of thirty-one syllables, by the poet Mokichi Saitō (1882-1953), which reads:

> With my mother dyin'
> In the tense silence
> I hear the frogs on paddies in the distance singin'
> Their songs echoing in the celestial sphere.

This verse captures the striking contrast between the stifling, silent and depressed atmosphere which hung over his mother's deathbed and the pulsation of life symbolized by the croaking of the frogs.

KIGUCHI: I once read the poem 'Mother' by Mr. Ikeda, and was moved by the following verse:

> Only the sympathetic resonance
> Between mother and child

   Reflects the beautiful and profound affection
   Inherent in all human beings.

SHIMURA: In the verse before this one by Mokichi, he superbly captures the moment which borders between life and death, wavering back and forth above his mother's bed, the light flickering in his eyes.

IKEDA: The sensitivity of the poet who aspired to create inspiring literature could not ignore the profundity of the shift from life to death which he had witnessed in a moment of high solemnity.

KIGUCHI: As a physician also, Mokichi must have been present at the bedside of some of his patients who were dying.

SHIMURA: Recently a physician published an account of the deaths of patients he had personally treated, a book which made the bestseller list. It is a report which concerns the cold reality of death. The author observes that even patients of high social standing who had been possessed of calmness and self-assurance became utterly terrified when it actually came to confronting death. Overtaken with fear, they would cry, 'Oh God, help me!' 'I don't want to die. I have so many things left to do,' etc.

KIGUCHI: Modern medicine is as yet very far from being able to alleviate in people the panic and fear of death.

SHIMURA: Regardless of how objectively one may have viewed death, an intellectual understanding does not allay his fears.

IKEDA: An objective approach to the problem of death based on scientific knowledge is one thing, but actually resolving it is quite another matter entirely. It is only one's own life or one's 'self' which can penetrate the seeming finality of death.

KIGUCHI: Only those who actually had brushes with death can

know its terror and fear. There are many people who had the experience of narrowly escaping death. For example, a man who was sentenced to death survived because the gun of the executioner misfired. At that moment, the prisoner's hair suddenly turned white.

IKEDA: The Russian author Dostoyevsky had a similar experience. It became a turning point in his life, and from then on, using that experience as an impetus, he went on to create great literary works.

SHIMURA: I am reminded of a poem written by Mr. Ikeda, 'Life of a Literary Giant', which sheds light on Dostoyevsky's works as well as on his thoughts and personality.

As I recall, Dostoyevsky was arrested when he was young for allegedly having engaged in subversive political activities. He was accused of reading aloud in public a work of a renowned literary critic, Vissarion Belinsky, entitled *Letter to Gogol*, which was banned in Russia. Dostoyevsky, considered a dangerous political element, was sentenced to death.

IKEDA: Dostoyevsky and some other prisoners were to be executed. The first three were each bound to a stake, and at the very moment that they were to be shot, an order to stop the execution arrived from the czar. A commissioner waved a white handkerchief in order to relay the message. In that brief moment, the moment that they were to have died, one of the prisoners went mad instantly, and another's hair suddenly turned white. Dostoyevsky was among the third group of prisoners and later recalled that his shock was beyond the power of words to describe.

SHIMURA: However, as it turned out, it had been a hoax all along. The execution was systematically arranged and then cancelled in order to demonstrate the czar's benevolence.

IKEDA: Those in authority at that time were using young men to secure their own power. But even today such things go on, though the circumstances may be different. Scheming by those in authority stems from the diabolical nature of power. In any event, Dostoyevsky wrote a great deal about the evil that festers within power and the anxieties and sufferings of human beings who could not penetrate the mystery of death.

KIGUCHI: Such works as *Crime and Punishment*, *The Idiot* and *The Possessed* reveal his genius for depicting the dilemma of human existence.

SHIMURA: Mr. Ikeda's poem about Dostoyevsky made me realize that the writer not only experienced bitter struggles during his life, but also immersed himself among the masses.

IKEDA: I don't believe that it was just his imagination that enabled him to write about the problems of life and death.

SHIMURA: Dostoyevsky knew all too well that the human mind would never be able to truly understand the mystery of death. This is something that goes beyond anyone's imagination.

Buddhism's View of the Moment of Death

IKEDA: In Buddhism, the trauma that one might experience at the moment of death is related in Nichiren Daishōnin's writing, *Jūō santan shō*, (*The Praise of the Ten Kings*): 'When one's life span comes to an end and he approaches the gates of death, as his final agony, various illnesses stemming from the eighty-four thousand illusions arise and vie in tormenting him. It is as though his body were being hacked to pieces by a hundred thousand halberds and swords. Therefore darkness fills his eyes, and what he wishes to see, he cannot; his tongue fails him, and what he wishes to say, he cannot.'

This writing goes on to state: 'Moreover, in the *Shōgon ron* it is stated that when one's life comes to an end, he perceives utter darkness, and it is as though he were falling into an abyss; he travels alone through a wilderness with no one to bear him company. Truly, at the time when one's spirit departs, his eyes are filled with darkness, and it is as though he were falling from a high place into great depths. When one dies, he wanders all alone through a vast plain. This is called the journey between death and rebirth.'

A person who dies in such a manner cannot express the anguish he experiences; others can only perceive his suffering by the pained expression on his face. The people at his bedside can only guess, but never really understand, the anguish of death. Only those people who have had brushes with death have any idea. Wealth, honour and social position are no match for death. This is the harsh reality of life.

SHIMURA: Thinkers or philosophers who attempt to probe the true nature of death usually formulate their own ideas about it.

IKEDA: Dr. Arnold Toynbee, in referring to the fact that in reality we are always in the process of dying, said in *Man's Concern with Death*, 'From the moment of birth there is the constant possibility that a human being may die at any moment; and inevitably this possibility is going to become an accomplished fact sooner or later. Ideally, every human being ought to live each passing moment of his life as if the next moment were going to be his last.'

However, he recognized that living on this spiritual level is indeed very difficult. Nevertheless, he says, 'What can be said with assurance is that, the closer a human being can come to attaining this ideal state of heart and mind, the better and happier he or she will be.'

SHIMURA: His deep insight is admirable. I once had a talk with the literary critic Hideo Kobayashi, and when he spoke about the resolution to the question of death, he said, 'It is very difficult for most people to comprehend death, so they must first seek out those people who know and understand human beings or those who are familiar with the question of death.'

KIGUCHI: It is virtually impossible for anyone to perceive the true nature of death within a limited, finite period of a given lifetime.

IKEDA: Shakyamuni's decision to renounce secular life was the result of *shimon yūkan*, or the four meetings. As a young prince, he lived the life of royalty, but one day, it is said, he exited from the eastern gate of the palace and encountered a man withered with old age. On another occasion, at the southern gate, he saw a sick person. At another time, he saw a corpse from the western gate. Finally, he emerged from the northern gate and encountered an ascetic. These events are the so-called 'four meetings' which motivated him to renounce the secular world in order to resolve the problem of the four sufferings, that is, birth, old age, sickness and death. After undergoing ascetic practices and meditation for a long period, Shakyamuni achieved enlightenment and in so doing realized the way to become emancipated from the sufferings of birth and death. He vowed to save mankind by revealing the truth to which he became enlightened. This truth is revealed in the *Lotus Sutra*.

SHIMURA: You have just now touched on the problem of birth and death and its resolution, based on Shakyamuni's enlightenment. What does the Buddhism of the Latter Day of the Law – that is, the present era – expound concerning the Law that enables one to free himself from the sufferings of birth and death?

IKEDA: In Nichiren Daishōnin's writing *Shōji ichidaiji kechimyaku shō* (*Heritage of the Ultimate Law of Life*), we find the passage: '... the life and death of all phenomena are simply the two phases of *Myōhō-renge-kyō*.' This passage points to the fundamental Law, which pervades not only human existence but all phenomena as well, and which reveals itself in the two phases of life and death.

KIGUCHI: This fundamental Law pervading both life and death is, then, the absolute reality?

IKEDA: That's right. It is the supreme principle spanning the eternity of past, present and future. To state it more clearly, it is *Nam-myōhō-renge-kyō*. This Mystic Law is itself the 'heritage unbroken since the infinite past', to use the words of the Daishōnin in the above-mentioned writing.

KIGUCHI: Then, we must conclude that, regardless of one's personal feelings or opinions, unless he bases his life on the Mystic Law, he cannot arrive at a fundamental or complete solution to the problem of birth and death.

SHIMURA: If so, then what does the Buddhism of the Latter Day of the Law teach concerning the problem of birth and death?

IKEDA: To explain simply, it is generally thought that life begins with birth and ends with death. However, Nichiren Daishōnin taught that life is eternal, spanning past, present and future with neither beginning nor end. Birth and death are the two inherent phases which one's eternal life repeatedly manifests.

With respect to the concept of eternal life in Nichiren Daishōnin's Buddhism, the *Ongi kuden* states, 'We repeat the cycle of birth and death secure upon the soil of our intrinsic enlightened nature.' It also states, 'Our birth and death are not the birth and death that we experience for the first time, but the birth and death that are forever inherent in life.'

'*Myō* Represents Death, and *Hō* Represents Life'

IKEDA: The *Shōji ichidaiji kechimyaku shō* also states, '*Myō* represents death, and *hō* represents life.' To explain briefly, the workings of birth and death, which are eternally inherent in life, manifest the Mystic Law as their true entity. When an individual life manifests itself in concrete form, that is called 'birth', and when it dissolves back into a latent, invisible state, it is called 'death'.

SHIMURA: Even if one is told, for example, that life 'melts' back into the universe after death, he can neither conceptualize nor sense it; therefore, death is represented by *myō* or mystic. Is this interpretation correct?

IKEDA: Yes. Simply speaking, we may consider this to be correct. Also, Dengyō states, 'Birth and death are the mysterious workings of the life-essence.' Again, in simple terms, that which activates and moves the individual life is called *myō*, the life-essence. When an individual life becomes completely exhausted, it enters the state of death in order to rest and be restored. The *Ongi kuden* states, ' "To depart" means to open out into the universe.' As this passage indicates, 'departure', or death, is itself a merging of the individual life into the great cosmic life. Then, by virtue of *myō*, the essential power or life force of the cosmos, it is recharged, so to speak, and born into the world anew. The interval of latency is called 'death'.

KIGUCHI: You said that *myō* is the motive force which activates life. What does that mean?

The Three Meanings of Myō

IKEDA: In the *Hokekyō daimoku shō* (*The Daimoku of the Lotus Sutra*), Nichiren Daishōnin explains this as part of the 'three meanings of *myō*'. They are: to open, to be perfectly endowed, and to revive.

KIGUCHI: What does 'to open' mean?

IKEDA: It is something like the power to open the door to a room in which boundless potential is stored. In short, when one believes in and practises the Mystic Law, his life begins to pulsate with vitality, and he can manifest the fundamental life of the cosmos as his own life force. It is said that the universe is expanding without limit; this, too, I feel, might be said to correspond to the meaning of 'to open'.

SHIMURA: What about 'perfectly endowed'?

IKEDA: Take, for example, a single drop of seawater: it contains the same constituents and qualities as the vast ocean. In an analogous way, each entity of life in the cosmos is endowed with all the properties of the cosmos itself. I believe 'perfectly endowed' can be understood in this way.

KIGUCHI: In a similar way, we might say that a rock brought back from the Moon by American astronauts contains the components of the Moon itself. Herein lies a fundamental premise of science, which seeks to clarify the universe. In even the smallest area of space around us, the same principles and forces are operating as in the great universe itself. This forms the starting point of Einstein's general and special theories of relativity.

IKEDA: Moreover, concerning the meaning of 'perfectly endowed', Miao-lo's *Maka shikan bugyōden guketsu* (*Annotations on the Maka shikan*) states, 'Because it can cure that which is thought to be incurable, it is called *myō* or mystic.' That is, 'perfectly endowed' also means that the Mystic Law possesses the unfathomable and boundless power to transform even the most anguished, tormented life into one of hope and joy.

SHIMURA: The passage '*Myō* means to revive', in the Daishōnin's writing, is quite significant, isn't it?

IKEDA: 'Revive' here means to return to life – that is, for a life to go from a state in which it has lost its vital energy to one in which it brims with fundamental life force.

KIGUCHI: This principle carries tremendous hope for people who feel sorrow over their destiny, and for humanity as a whole, which has reached a stalemate.

IKEDA: 'To revive' means that, by embracing the Mystic Law, each person without exception can carry out a vibrant and boundless reformation of his or her life itself. This reformation forms the basis for the perfection of one's character and, moreover, provides the power to transform one's destiny.

Herein lies the significance of achieving 'human revolution', a concept proposed by the late Shigeru Nanbara (1889-1974), a political scientist and former president of Tokyo University. I also believe that our purpose and mission is to contribute to the community, society and to the world, and to respond to the needs of the times on the basis of this 'human revolution'.

KIGUCHI: In other words, this means the individual's sense of purpose, based on the principle of taking action within society as well as making an effort toward the realization of peace, with the teachings of Buddhism as the foundation.

SHIMURA: Professor Nanbara proposed the concept of 'human revolution' at a graduation ceremony in 1947, but did not put it into concrete terms at that time. The Japanese writer, Sōhachi Yamaoka (1907-1978) also used the term 'human revolution'.

IKEDA: It is apparent that in the innermost depths of their being, people long to live a value-creating life; the unlimited creative potential originates within the human being himself. For this reason, man is earnestly seeking the Law which is limitless so that he can manifest this infinite creative potential.

KIGUCHI: This is the underlying foundation of scholastic and scientific pursuit, as well as other endeavours such as business ventures.

IKEDA: In this sense, if we view it objectively, our religious activities exist in the dimension of a great movement which can revive life itself, the basis of all our endeavours.

KIGUCHI: The Buddhist view of life and death emanates from a particular vantage point and expands to fill the entire universe.

SHIMURA: The intelligentsia are now beginning to recognize the need to approach the question of life and death from many perspectives.

The Possibility of Terrestrial Life on Other Planets

SHIMURA: This year (1983), there have been various news topics in the field of astronomy. What interests you the most, Dr. Kiguchi?

KIGUCHI: Well, Dr. Cyril Ponnamperuma's research which provides evidence of the possibility of terrestrial life on other planets is fascinating. The nature of his research is somewhat difficult to explain, but it has some connection with the work of Dr. Chūshirō Hayashi, the scholar under whom I am working.

IKEDA: In what way are they related?

KIGUCHI: Dr. Hayashi is now investigating the birth of our planet, and his explanation is considered plausible by most scientists. His theory is derived from the nebular hypothesis set forth by French astronomer and mathematician Pierre-Simon Laplace (1749-1827). Briefly, Laplace theorized that a hot gaseous nebula gradually cooled and contracted and through this process the solar system evolved. During his lifetime, however, Laplace had little data to support his hypothesis, so it remained unsubstantiated. These days, however, there is a tremendous amount of data available because of the interstellar explorations being undertaken. Therefore, we are now in a position to verify the most important points of this theory. It is postulated that in its earliest phase our planet consisted of a cloud of gas and dust. Small particles in the cloud collided with each other to form larger particles. These conglomerates became so large that they drew gaseous dust particles from space due to gravitational attraction. The kinetic energy of the smaller particles' collision with the larger ones was converted into heat. Heat was also produced by the decay of radioactive isotopes in the Earth. The temperature of the planet, therefore, increased, and matter in the Earth fused together. As a result, the heavier elements sunk into the planet's core while lighter elements rose to its surface; thus a separation between the inner and outer zones of the planet took place.

IKEDA: So the Earth is thought to have evolved in this manner. The planet's core and its mantle were formed in this way, is that correct?

KIGUCHI: That's right. When the crust of the Earth cooled, water vapour in the atmosphere condensed and formed the oceans. In the meantime, there is evidence to indicate that, in the early stages of the planet's development, a large number of ions were present in the atmosphere, resulting in frequent electrical storms; also many meteors struck the planet. In the midst of all this chaos, however, it is thought that the environment became gradually suitable for the generation of life.

Dr. Ponnamperuma's experiment simulated the atmosphere at the time of the Earth's early evolution. He succeeded in simultaneously producing all five types of bases essential to life's genetic code by passing an electrical charge through the gas of the simulated atmosphere.

SHIMURA: The results of this experiment have tremendous implications for the possibility of the existence of terrestrial life elsewhere in the universe.

IKEDA: What was solar ultra-violet radiation like then, that it would have been harmful to living creatures?

KIGUCHI: Because of the absence of ozone in the Earth's atmosphere, ultra-violet rays penetrated the ocean to a depth of ten metres below the surface. During the Earth's primordial stage, only a small amount of ozone was formed as a result of photodissociation. Through the action of photosynthesis, atmospheric oxygen increased. As a result, the amount of ozone has increased proportionately. This ozone layer thereby became dense enough to block out the harmful effects of ultra-violet radiation. Thus, the condition for the proliferation of life on Earth was set. In the meantime, the photosynthetic process resulted in a steady increase in the amount of free oxygen, which amounted to about one-tenth the volume of oxygen which exists today, thereby enabling a blanket of ozone to form and protect the Earth from ultra-violet rays.

IKEDA: This is indeed an example of the great wonder of nature. One can only call it mysterious. I am overwhelmed with awe. A lay person such as myself can conjure up no words of admiration except possibly 'wondrous'. So all the necessary conditions for life to flourish on this planet thus took shape.

SHIMURA: Within the vast universe, a seemingly unending void, our precious planet exists. Numerous forces, all interrelated, are factors which generate and support life on Earth. The more I think about this, the more awestruck I become.

KIGUCHI: I feel the same way. However, in the distant future, the Earth and the Sun will disintegrate.

IKEDA: This brings us to the phenomenon of the life and death of the planet Earth itself. All phenomena in the universe, whether living beings or matter, are subject to the strict laws of nature.

KIGUCHI: It is this truth that has led astronomers to discover that even when our solar system dies, this is not the end of the story. After an extremely long period, a new sun will be born and new planets will come into existence in the wake of the destruction of the previous solar system.

SHIMURA: Does this mean that a planetary system similar to our own will again evolve?

KIGUCHI: That's right. There are, in fact, some scientists who also conjecture that in this new solar system, intelligent life similar to human beings will evolve.

A number of astronomers consider the present research into the formation of planets to be one of the greatest achievements of modern science. And it is no small wonder that these findings concur with the views of the universe held by Buddhism which espouses, among other things, the concept of all phenomena constantly repeating the cycle of birth and death.

From the standpoint of chemistry, we find that on Earth, the soil, plants, animals and human beings all originate from the atoms which comprised the universe during its early stages. The other planets, stars and galaxies which presently exist are said to have evolved after the Big Bang.

Dr. Robert Jastrow, the American physicist and author, says that in this great universe, both sentient and insentient beings are closely interrelated. He also expresses his marvel at the striking similarity between the idea of the emergence of the heavens and earth and life as described by astronomers and the theory of how life evolved as expounded by Buddhists and philosophers of the East.

IKEDA: Buddhism views the living being and its environment as being inseparable. On this point, Nichiren Daishōnin states that, while life and its environment are two independent phenomena, they are one in their fundamental essence. Therefore the individual entity is shaped and influenced by the environment, and, in turn, the individual exerts an influence on the environment. My mentor, Jōsei Toda, often remarked, 'A true religion neither runs counter to, nor conflicts with, science.'

KIGUCHI: Although it is unlikely that Dr. Jastrow had a complete grasp of the entire system of Buddhist thought, I believe that the revolutionary discovery that the ultimate origin of mankind was indeed a cloud of gases in a small sector of the universe led him to this conclusion.

IX
The Life of a Star

How does the movement of the universe relate to human activities? How is a star born and how does it die?

The Dawn of the Space Age

SHIMURA: A recent topic in the news is Japan's intention to begin recruiting her own astronauts, in cooperation with the US space shuttle programme.

IKEDA: How many people do they want?

KIGUCHI: Three, but, of course, many people will no doubt apply.

SHIMURA: Men and women can apply, but only people who have had some sort of scientific or technological training with a minimum of five years' experience will be considered.

KIGUCHI: Five years after they have been selected and trained, one will be carried aloft as a member of the crew on one of the US space shuttle missions.

IKEDA: The first man in space was the Soviet Union's Yuri Gagarin in the *Vostok 1* – about twenty-three years ago. Space has become the symbolic backdrop of the twentieth century. It was during this century that nuclear power and nuclear weapons first came into use, and that the space age began. The human perspective has now become magnified to a global scale. The point is that we can no longer discuss peace and freedom from the limited viewpoint of national interests.

It gives me tremendous pleasure to see some of the science fiction which I read as a boy becoming a reality. It is my hope that this great leap in human ingenuity will be used to realize world peace in addition to further developing scientific technology. When was the possibility of space travel seriously considered?

KIGUCHI: Around the 1920s.

SHIMURA: About sixty years ago a professor by the name of Robert H. Goddard of Clark University wrote an article in a scientific journal in which he said that in the future it would be possible for rockets to reach the Moon. It caused quite a sensation.

KIGUCHI: Professor Goddard is an American physicist who came to be called the 'Father of Rocketry'.

SHIMURA: As a matter of fact, at that time, *The New York Times* criticized him severely, declaring that his idea was an impossibility and that the professor lacked sufficient knowledge concerning the vacuum-like state of outer space.

IKEDA: In every age, people of foresight are criticized and despised by the public. However, half a century later, the spaceship *Apollo 11*, manned by Astronaut Neil Armstrong and others, succeeded in landing on the Moon.

SHIMURA: That was in 1969 and, moreover, *The New York Times* published an official apology to the late Professor Goddard. Other journalists around the world considered this incident to be a good example of journalistic conscience.

KIGUCHI: It must have taken courage.

IKEDA: Moreover, since it has a global readership, the decision to apologize must have been a difficult one to make.

KIGUCHI: A space centre in the United States was named after Professor Goddard.

IKEDA: That's a great story.

KIGUCHI: Professor Goddard was a great scientist, and *The New York Times* showed integrity.

IKEDA: It seems to me that that newspaper has always attempted to present a broad view with a sense of humility. Their desire to keep the confidence of the public and maintain their credibility was well demonstrated in this instance.

The Significance of New Year's Day

SHIMURA: What is the significance of New Year's Day with regard to the movement of the heavenly bodies?

KIGUCHI: In ancient times, people determined the first day of the year from their observations of the heavenly bodies.

IKEDA: I read that in ancient Egypt, New Year's Day fell on the autumnal equinox, while in Judea and Babylonia, it was celebrated during the spring equinox; in Greece it was at the winter solstice.

KIGUCHI: In Japan, New Year's Day fell on the first day of the first month of the lunar calendar,[1] which is somewhere between 21 January and 19 February according to the Western calendar.

SHIMURA: It is interesting that the season for New Year's Day differed according to the culture. In Japan, we call the morning of New Year's Day *gantan*, which means 'the first dawn'.

IKEDA: Actually, the character *gan* means 'beginning' and *tan* symbolizes the Sun rising above the horizon.

SHIMURA: The Japanese custom of writing the characters

gantan on traditional New Year's greeting cards reflects the importance placed on the first sunrise of the year.

IKEDA: Both characters, *gan* and *tan*, originated in China, of course, but it is also true that from very early times, people's daily lives were very closely related to the movement of the Sun, in both Japan and China.

KIGUCHI: The ancient Chinese farmers celebrated the rising of the Sun on New Year's Day.

IKEDA: Perhaps they clearly delineated the division of time within their day-to-day activities in order to maintain a fresh feeling of renewal. In other words, they wanted to preserve the feeling of renewed joy and appreciation for life by being aware of the rhythm of the heavens. Their celebration of New Year's Day is one such expression of this desire.

KIGUCHI: I have friends who insist that they are too busy studying the stars to take any notice of the *Bon* festival[2] in August or of the end of the year, but on New Year's Day they take stock of themselves, wear a happy smile and even don a Japanese kimono.

SHIMURA: Perhaps due to the influence of television and mass communication, people seem to take an interest in the latest fashions and try to keep up with the latest developments in the news, but nevertheless they still value old-fashioned traditions.

IKEDA: People are not satisfied with just knowing about the recent events; rather, they seem to seek out something which is lasting – a way to sense eternity, so to speak. Ultimately they seek peace of mind, a sense of fulfilment and complete satisfaction within their hearts.

KIGUCHI: The human intellect tries to grasp a feeling of permanence.

SHIMURA: People who do not ordinarily go to temples or

shrines suddenly become 'religious' and go on New Year's Day. However, there is a difference between doing what is fashionable and respecting tradition.

KIGUCHI: By the way, how do you spend New Year's Day, Mr. Ikeda?

IKEDA: In my case, the New Year's celebration actually concludes on New Year's Eve. It is our family custom to gather at midnight and conduct *gongyō* together at the family Buddhist altar. Following that, I pour sake for everyone and we eat the traditional New Year's dishes. On New Year's Day, I eat breakfast as usual and begin the year's activities.

My birthday is 2 January which is one of the New Year's holidays. For the last few decades, I have made it a custom on that day to make a pilgrimage to the Nichiren Shōshū Head Temple, Taiseki-ji, located at the foot of Mount Fuji. For me, the New Year season may be the busiest, but I try to begin each day with the same refreshed feeling that I have on New Year's Day.

KIGUCHI: An old Chinese proverb says 'Begin each day anew.'

SHIMURA: Various customs have evolved for celebrating New Year's Day and have been handed down for many generations. Each of them has a story about its origin.

KIGUCHI: In Osaka, where I live, it is said that in the old days people did not put up pine trees at the front gate of their homes, although this is now a common Japanese New Year's tradition.

IKEDA: It isn't done at the Imperial Palace either, so it can't be an old custom handed down from ancient times.

SHIMURA: According to one explanation, in the Meiji period (1868-1912), the Ministry of Education apparently promoted a song which went, 'Let's put up pine trees at every gate to make our celebration of the New Year even merrier.' Because the song became popular, the custom spread throughout the country.

IKEDA: The sending of New Year's cards has its origin in the visits traditionally made during New Year. The postal system was well established by about the middle of the Meiji period, and it became fashionable to send name cards in envelopes. It then became the custom to send New Year's cards.

KIGUCHI: I enjoy receiving New Year's cards written in traditional Chinese calligraphy by friends whom I am not able to see often, although I am given to writing mine with only a fountain pen.

SHIMURA: I often see you with a brush in your hand, Mr. Ikeda.

IKEDA: I'm an amateur calligrapher. I do it because my friends urge me to, but I sometimes feel that the results may prove to be embarrassing in the future. I only do it to make them happy and to encourage them.

SHIMURA: A professional calligrapher must study for many years to strengthen his hand enough to be able to write thousands of characters. This takes almost half a lifetime, and probably more than that to be able to move one's hand firmly and evenly. Your calligraphy expresses your heart and reveals your dynamic life force. It must be difficult to find time out of your busy schedule.

IKEDA: I've noticed that I can move the brush the way I want to on a fine day, but not very well on a rainy day. The result depends on the weather and on my state of mind.

KIGUCHI: It is often said that one's handwriting reveals one's character. Calligraphy which is imbued with the writer's sincerity and humanity is attractive. It doesn't matter whether it is executed according to strict artistic standards.

IKEDA: I agree, but not because I have a poor hand.

SHIMURA: Calligraphy is one of the fine arts. The famous sculptor Auguste Rodin (1840-1917) said, in effect, that creativity is not necessary for art, but life is. This quality is the essence of art.

IKEDA: Nichikan Shōnin, paraphrasing T'ien-t'ai's *Maka shikan*, refers in his *Sanjū hiden shō* to '... an excellent painting reproducing an image in a way that is truly compelling, in form and spirit vivid and alive.'

KIGUCHI: Whether in a painting, calligraphy or through music, a genius seems to be able to skilfully express his or her outlook on life, the world or even the universe.

SHIMURA: I attended the exhibition of modern French painters at the Tokyo Fuji Art Museum. In viewing *Hellas Mourning over the Ruins of Missolonghi* by Delacroix, I realized that the painting is nothing less than a masterpiece. I was overwhelmed by this work and the other masterpieces there.

IKEDA: Works which inspire in the viewer such feelings which you just described are art of ultimate aesthetic beauty. Artists of genius understand their subjects fully through their mind's eyes, and then create their works just as they envisioned them.

KIGUCHI: The human mind is deeper than any ocean and wider than the sky.

SHIMURA: My favorite English poet, William Blake, wrote a profoundly lyrical poem:

> To see a world in a grain of sand,
> And a Heaven in a wild flower,
> Hold Infinity in the palm of your hand,
> And Eternity in an hour.
> ('Auguries of Innocence')

IKEDA: Henri Bergson, Albert Einstein and others are said to have appreciated this poem because it captures the relationship between the human being and the universe. Blake must have perceived the expanse of the entire universe in one individual. Perhaps we could call this a 'poem of life'. As I have often said, Buddhism clearly explains the relationship between the 'self', that is, the inner universe, and the outer cosmos. Nichiren Daishōnin said, 'In the final analysis, all phenomena are contained within one's own life, even down to the last particle of dust. The nine great mountain ranges and the eight vast seas[3] are contained within one's body, and the sun, the moon and the stars are contained within one's life.'

The Intercalary Month

SHIMURA: This may not be directly relevant to our discussion, but, Mr. Kiguchi, would you please explain why February is sometimes referred to as an *intercalary* month?

KIGUCHI: According to an old Roman calendar, March, now the third month, marked the beginning of the year. Therefore, during leap year, an extra day was added at the end of the year, that is, at the end of February. That custom has been carried over into our present system (the Gregorian calendar).

IKEDA: The origin of the character for 'intercalary' is most interesting. According to Chinese dictionaries and other sources, the emperor would visit his ancestral temple to preside over the ceremony heralding the phase of the New Moon. During the extra intercalary month, however, he remained within his own gates.

KIGUCHI: In the past, the ruler would visit his ancestral temple on every first day of the month.

IKEDA: During an intercalary month, however, he did not go beyond the gates of his residence. Hence the character for 'intercalary' is written with the symbol for 'ruler' inside the radical representing 'gate'.

KIGUCHI: At that time, the Chinese went by the lunar calendar, according to which one month has twenty-nine days. This leaves more than ten extra days each year. Therefore, about every three or five years – a total of seven times in nineteen years – they would add an 'intercalary month' to make up for the discrepancy between the lunar calendar and the seasons, determined by the Sun.

SHIMURA: I understand that at that time, Japan did not as yet have a set calendar.

IKEDA: I researched this topic, and the *Wei chih*[4] (*Wei zhi; History of the Kingdom of Wei*) states, 'The people of Wo (an ancient Chinese name for Japan) have no calendar, but simply divide the year according to the spring sowing and autumn harvest.'

SHIMURA: The original Japanese name for calendar – *kayomi*,[5] literally 'reading the Sun' – comes from counting the number of times that the Sun rose during the course of the year.

KIGUCHI: Even in ancient times, the Chinese had a very advanced calendar.

IKEDA: Actually, the Chinese calendar reflects strong Buddhist influences. It incorporates a number of ideas presented in the *Matōga, Nijūhasshuku, Daishutsu, Shukuyō* and other sutras, which were introduced from India.

SHIMURA: The ancient Indians had an excellent calendar which appears to have been influenced by Buddhist cosmology.

KIGUCHI: That stands to reason. The Chinese in turn applied these ideas and, by the eighth century, developed the most advanced calendar in the world.

SHIMURA: That was when the *Ta yen li* (*Da yan li; Great Developed Calendar*) appeared, a work based on Buddhist cosmology, detailing the movements of heavenly bodies. This work, in fifty-two fascicles, was introduced to Japan during the Nara period (710-794).

KIGUCHI: This work greatly impressed scientists worldwide because of its precise calculation methods. Granted, Buddhist cosmology may have been primarily intuitive and was not based on modern techniques used in scientific observation; but, interestingly enough, it coincides on a great many points with the findings of modern science.

The Sun and Earth Find Their Origins in Interstellar Matter

KIGUCHI: I'm now studying the formation of our solar system which is thought to have evolved from clouds of interstellar gases. In technical terms, this is involved in 'the problem of gravitational equilibrium'. Simply stated, we assume that the following process led to the formation of the Sun, Earth and the other planets: Interstellar gas and matter were travelling at great speed. As they revolved around a focal point, the nearby gases and particles in the surrounding space were pulled in. After billions of years, a disc-shaped cloud of gas and dust was formed. Then, because of gravitational forces, condensation began to take place at the disc's centre. After ten to twenty million years, the centre became a proto-star – a dense gaseous nebula about as large as our present Sun. At the same time, small clusters of matter were produced by local gravitational forces in the clouds of gas revolving around the proto-star. Within these gaseous clouds, these small clusters travelling at relatively slower speeds collided and adhered to each other. After about five thousand years, these conglomerations also formed gaseous disks of their own, which coalesced into grain-sized particles in the amazingly short period of one year. These materials stuck together, moving about at relatively slow speeds, and formed larger conglomerates about twenty kilometres in diameter. This process continued until the Earth and the other planets were formed.

I am now using computer simulation to determine how the Sun and its planets were formed by this disc-shaped cloud of gas and dust during the unimaginably long period of the solar system's evolution.

IKEDA: This study is fascinating. We can't see interstellar matter such as gas and dust, can we?

KIGUCHI: With the exception of dark nebulae, no, we can't. The invisible gas and dust, which are the fundamental components of our solar system, are perhaps the remnants of an ordinary star that once existed. When the star exploded during its final moments, it released tremendous amounts of energy into space.

IKEDA: Space seems to be a mere void, but actually it isn't. Its composition is extremely complex. The seeming void of space contains abundant matter just as our bodies contain bacteria which are invisible to the naked eye. The study of this invisible entity is indeed important and fascinating.

KIGUCHI: What interests me most about theoretical astronomy is that through computer simulation we can model the evolution of heavenly bodies over a period of billions of years.

SHIMURA: Dr. Kiguchi, you mentioned that the disc-shaped clouds of gas and dust further condense and congeal into matter the size of grains of sand within one year – an extremely short period in terms of astronomical time. Since the evolutionary process took millions or billions of years up until then, how were you able to determine that this process took place within such a short time?

KIGUCHI: That's one of the key points in the theory proposed by Dr. Chūshirō Hayashi. A few years ago, he made his calculations and announced the results to the laboratory staff members. It would take pages to explain his methods, but suffice it to say, one year was arrived at.

SHIMURA: It seems to me that the stars and the human body are essentially composed of the same things. In Buddhism, everything is understood to arise from the harmonious fusion of the four elements — earth, water, fire and wind. Thus, the force that caused interstellar matter to condense can be thought of as 'moisture'.

IKEDA: According to a Buddhist treatise,[6] the essence of water among the four elements is moistness, and its function is to gather and hold together other matter.

SHIMURA: By the way, how does an infra-red telescope work?

KIGUCHI: Infra-red rays are emitted by relatively cool bodies. These rays have strong penetrability so that an infra-red telescope can be used to detect distant, relatively dark objects which cannot be seen with an ordinary telescope. Thanks to this device, we can now observe objects in space far more accurately than ever before.

IKEDA: I recall an article in the *Yomiuri Shinbun*[7] about making observations from the infra-red astronomical satellite (IRAS), a joint project involving three countries: the USA, the UK and the Netherlands.

SHIMURA: Yes, that's right. The article reported the discovery of ultra low-temperature celestial bodies in the minus 220° C range and three dust belts in the solar system. That satellite was launched in January 1983.

IKEDA: I was particularly interested because the existence of a second solar system in our galaxy was officially confirmed.

SHIMURA: As was mentioned earlier, this second solar system, discovered by the Jet Propulsion Laboratory in California, after their analysis of the IRAS data, exists around Vega in the constellation Lyra.

KIGUCHI: Its existence was again confirmed on 9 November 1983, and officially announced as the result of the findings made by the thirteen scientists from three countries involved in the joint satellite project.

SHIMURA: What process must a gaseous nebula undergo before becoming a star?

KIGUCHI: The most important factor is that the minute particles and gas must move very rapidly, thus converting gravitational energy into thermal energy, which increases in pressure within the mass. This results in a balance between the thermal energy which generates outward pressure and the gravitational energy which creates inward pressure.

IKEDA: Then the generation of thermal energy and the heat which results play an important role.

KIGUCHI: Yes. A star can't be formed properly until the heat spreads throughout the entire mass. The movement of the particles inside the gaseous nebula is very crucial at this time, because the shape and the location of the star to be formed depend on the quantity of particles and the nature of their movement. According to one theory, Jupiter was originally a gaseous nebula just like the Sun, but it is not a radiant body because it didn't acquire enough mass or gravitational energy to encourage the movement of particles.

IKEDA: The same Buddhist treatise[7] that I quoted before says that the essence of fire is heat and its function is to ripen. Accordingly, this energy and the heat you just mentioned are performing the same function which the treatise describes. Fire or heat is thought to be an important element that sustains human life. Therefore the treatise indicates that the life force sustains heat, in this case, body temperature, and consciousness; conversely, heat and consciousness sustain life force. All phenomena in space can be understood to conform to the universal principles expounded by Buddhism.

KIGUCHI: With regard to the balance between the outward expansion and inward contraction of a gaseous nebular or proto-star, if the movement of particles at the core is not very active, the thermal energy cannot be maintained, and so the temperature decreases and the outward pressure decreases. As a result, the development of the star is arrested at this stage.

IKEDA: I understand that energy which is generated from the centre is very important. I think this is true for a social organization as well, in which the great energy of its leader becomes the driving force of its movement. On the other hand, if the movement of particles at the core is too rapid, the energy and temperature of a proto-star become too extreme, and its activities will proceed too quickly, inhibiting the development of the star. Likewise, if the leader of an organization is too dogmatic, emotional or egotistic, the harmony of the organization as a whole is destroyed.

Whether in natural phenomena or human relationships, harmony can be maintained only in relation to the Mystic Law, 'the ultimate truth of the Middle Way'.

SHIMURA: You mentioned that newborn stars are called proto-stars or 'stars in their primitive stage'. Why do you refer to them this way?

KIGUCHI: Their activities have not stabilized or matured, so that the light which they emit fluctuates. Although the core has become stabilized, they are still affected by the remaining gaseous clouds and, for a while, are sensitive to their immediate surroundings.

IKEDA: This is also true of people. As long as one is easily influenced by his environment, his spiritual development is still in its primitive stage. Full development requires education, character development, good health, and other qualities. While the development of an independent self is very important, we must also realize that by contributing to society we can make our existence meaningful.

KIGUCHI: A proto-star evolves into a full-fledged star when its core temperature reaches 10^{7}° K. At this point hydrogen fusion takes place, producing tremendous amounts of energy. The fusion of just one gram of hydrogen releases as much energy as twenty tons of burning coal. This energy becomes the outward pressure which balances itself against the inward pressure created by the star's gravity. At that point, for the first time, the star stops contracting and begins to emit brilliant light.

IKEDA: It seems that motion can be calculated accurately throughout the entire universe. Although dust particles which ultimately become part of stars seem to move about freely, in fact they do not. They collide with one another and move away in a particular direction, and the same pattern is repeated until one unified body is gradually formed. I think that this depicts magnificently the drama of the birth of innumerable stars, based on the workings of the Law of the universe, *Nam-myōhō-renge-kyō*.

KIGUCHI: As a matter of fact, after this process has continued for a long time, a star suddenly appears. Our Sun began to shine in the blackness quite suddenly and a thousand times more brightly than it does now. It is quite surprising that the formation of this star which took billions of years culminated in a single moment when it suddenly burst into a huge ball of brilliant light.

IKEDA: That's fascinating. The repeated cycle of activity finally culminated in a tremendous transformation in one moment. Although stars are insentient and human beings are sentient, they have much in common in this respect. To give an example from history, Shakyamuni Buddha, after renouncing the secular world, spent more than ten years exerting unceasing effort to seek the truth of life and the universe, and finally attained enlightenment in one moment.

In Nichiren Daishōnin's Buddhism, there is the principle of attaining Buddhahood in one single lifetime through daily practice. (One lifetime is but a moment from the standpoint of eternal life.) In keeping with this principle, we should exert our maximum effort in pursuit of our goals. Then, and only then, can we manifest the 'self,' which is a miniature universe, within our lives. This is the drama of harmony, fulfilment and self-realization. On the contrary, people who do not consistently exert effort cannot expect to achieve anything remarkable.

SHIMURA: The dynamics of the great universe and the microcosm are identical.

IKEDA: From another point of view, the ultimate purpose of one's practice of Buddhism is the attainment of Buddhahood. It is a dimension which we cannot fathom with our intellect. As is taught in the sutras, Buddhahood is achieved through courageous actions, devoted practice and continuous self-reflection. If one fails to attain Buddhahood, or is unable to carry through his faith until the last moment, then one's practice of Buddhism becomes meaningless.

Inspiration Based on Deep Contemplation and Effort

KIGUCHI: I heard an interesting story about the pioneering work involved in constructing the theory of elementary particles by Dr. Hideki Yukawa. Day and night, he had been pondering for months the problem of elementary particles, and one day a major breakthrough suddenly happened. We must devote ourselves wholeheartedly to something and keep pursuing it in order to experience such moments of insight.

IKEDA: As a Buddhist, it is my conviction that the believer must do *gongyō* and propagate the ultimate teaching of Buddhism throughout the world. Faith, practice and study constitute the foundation, the bone and marrow, of this religion.

When I envision the Sun at the moment when light suddenly burst forth from the pitch-darkness of space, it reminds me of a king who appeared in order to send out the light of mercy.

There is a similar famous story about the discovery of the principle of buoyancy by Archimedes, isn't there?

KIGUCHI: Yes. One day he was ordered by Hieron II, king of the Greek city-state of Syracuse, to find out whether the royal crown was made of pure gold or not. Archimedes tried tenaciously for a long time to devise a method to determine this, but he couldn't find a solution. Then one day, when he was stepping into a public bath and observed water running over the side, he had a sudden inspiration. He was so overjoyed that he ran home without a stitch of clothing on, shouting, 'Eureka! eureka!' (I've found it, I've found it). Using the principle of buoyancy, he was able to calculate the specific gravity of the crown and found that the gold had been alloyed with silver. Thus, not only Archimedes but people of genius on the whole often seem to find inspiration in common occurrences which would pass unnoticed by others. Why does this happen?

IKEDA: There is a famous saying, 'Genius is another name for effort.' People of genius apparently have a 'flash of inspiration' out of nowhere, but their apprehension of the truth could never be considered coincidental. We must realize that their insight is the result of their long and continuous efforts.

SHIMURA: Most people are dreamers and want immediate results, but don't take the time to meditate on them deeply or exert enough effort to get what they want.

IKEDA: Those people who seek true meaning in life exert tremendous effort in an attempt to view the truth from new perspectives.

SHIMURA: They are possessed with a voracious curiosity, and this seems to be the case with such people in any field of endeavour.

IKEDA: During the course of history, as these people of genius sought out truth and discovered something of value, their efforts contributed to the well-being of mankind as a whole. Therefore, I think that we must live our lives with an open and penetrating mind, seeking the truth, unfettered by our emotions, which are transitory and biased.

SHIMURA: That is the true path of progress. But people today have a tendency to be more impressed by tangible material wealth. In this sense, ours is an unfortunate age.

Recently, the man who has discovered an artificial method of cultivating high-priced *matsutake* mushrooms made an interesting remark. It took sixteen years for him to arrive at his discovery, but he said in retrospect, 'Having come up with this method, I don't know what took me so long.'

IKEDA: This is not unusual. Similarly, a spiritual awakening may seem beyond the imagination of those who haven't experienced one. On the other hand, for those who have, it is an obvious truth.

SHIMURA: In other words, the joy of climbing to the top of a high mountain, for instance, can be understood only by those who have actually done it. Those who stay at the bottom of the mountain may say, 'It's so high that it must be very formidable.'

IKEDA: In a way, the real joy of living may be found in the effort one makes towards accomplishing one's goals. The greatest achievement in Buddhism is the revelation that the Law of life underlies all phenomena. Fusing one's life with this Law enables one to lead a harmonious and happy life.

The Lesson of Sessen Dōji

SHIMURA: As for meditation and practice in Buddhism, there is the famous story of Sessen Dōji related in the *Nirvana Sutra*.

IKEDA: This is said to be the story of Shakyamuni Buddha, in one of his previous lifetimes when he was a bodhisattva practising austerities as a Brahman.

It is written that, as Sessen Dōji, he had continued to contemplate the truth for innumerable years. One day, the god Taishaku (Skt Indra) appeared to him in the form of a demon, and taught him half a verse from a Buddhist text: 'All is changeable, nothing is constant. This is the law of birth and death.' Sessen Dōji was overjoyed to hear it and begged the demon to tell him the second half. The demon, however, told him that he was starving, and he would narrate the second half only if Dōji offered him his own flesh and blood. Dōji promised, and after being told the latter half of the verse which was: 'Extinguishing the cycle of birth and death, one enters the joy of nirvana', he quickly wrote it down everywhere so others could learn it, and then, from a tall tree, he threw himself into the demon's mouth. At that moment, the horrible demon was transformed back into Taishaku and caught Dōji before he fell. He praised him for his seeking spirit and willingness to pursue the truth or the Law without regard for his own life, and predicted that Dōji would no doubt become a Buddha.

KIGUCHI: That story illustrates the importance of discipline and practice in seeking the truth.

IKEDA: Buddhism isn't just theoretical. Even if one is well-versed in the Buddhist concepts and can expound them fluently, one cannot attain Buddhahood, the ultimate goal of the practice and the discovery of one's true 'self', unless one practises it strictly.

The Last Stage of a Star's Life

SHIMURA: Stars and planets like the Sun and the Earth, which were formed over a period of billions of years, will inevitably collapse and burn out. Could you tell us a little about this process?

KIGUCHI: Some stars become cold and contract, while others are dispersed into the universe.

IKEDA: Why do stars disintegrate in different ways?

KIGUCHI: It depends on their size. Simply speaking, stars smaller than the size of our Sun become colder and gradually contract. These are called white dwarfs. Stars larger than the Sun cannot maintain their shape, so they swell and then contract. Finally they explode and are dispersed into the universe.

The core of some of these stars remains after the explosion, and the remaining material forms a neutron star. Neutron stars are one of my favourite subjects, and I have studied them in great detail. It is thought that they will also be dispersed into the universe finally, after an enormous span of time – in years, something on the order of one followed by ten trillion times ten trillion zeroes.

IKEDA: The standard of comparison is our Sun. Does this standard apply to every star?

KIGUCHI: Yes. Our Sun is considered an average star.

SHIMURA: Is quantum mechanics an integral part of the study of the death of stars?

KIGUCHI: Quantum mechanics is the most appropriate discipline for studying and eventually arriving at a true understanding about the death of stars. I am presently investigating the movements and changes in elementary particles, atoms and molecules which make up the stars, from their birth to their death. For this study, only paper, pencils and sometimes a computer are necessary; it costs almost nothing.

IKEDA: Then am I to understand that quantum mechanics is a science which can explain the death of stars by studying invisible minute matter such as molecules, atoms and elementary particles?

KIGUCHI: Quantum mechanics is a science which explains how the materials of dead stars disintegrate into the expanse of time and space. When a star is dead, we cannot see it or perceive it under ordinary circumstances. As I mentioned before, neutron stars can tell us clearly, through quantum mechanics, how the unseen world of matter functions.

SHIMURA: How is this possible?

KIGUCHI: Roughly speaking, a neutron star is like a cinder which remains after a star's explosion. It does not shine, nor is it a source of energy. A star is a star only when it continuously creates outward pressure, caused by the thermal energy it generates. Therefore, if a star has no thermal energy, it cannot resist the pressure of gravity and it collapses upon itself and contracts to a miniscule size, to the extent that it can no longer be detected. However, neutron stars, which have no source of energy but are a certain size, have been observed in space, and were a mystery until about the 1930s. There is no way to explain their existence without taking into consideration the possibility that they must be influenced by an unknown force which exists everywhere in space and has no material substance, which can only be within the realm of the unseen world. However, we can detect and calculate the magnitude of this force through quantum mechanics.

IKEDA: This may be on another level, but it seems to be related to the idea that even when our bodies disintegrate, life itself or the 'self' continues to exist.

Just as I mentioned before, with regard to the Mystic Law or *myōhō*, Nichiren Daishōnin stated '*myō* represents death, and *hō* represents life.' This means that *myō*, though invisible, is dynamically manifested as *hō* which represents the visible phenomenal world. Therefore, as the latest achievements of scientific endeavour unfold, we can also come to understand more clearly the Buddhist principle that '*myō* represents death, and *hō* represents life.'

KIGUCHI: Through the development of quantum mechanics, phenomena which humans couldn't even have imagined have been verified.

IKEDA: The *Ongi kuden* states, 'The universe is *myōhō*.' The most fundamental concept among the Daishōnin's teachings is that if one prays to the Mystic Law, the ultimate Law of the universe, his prayer penetrates all aspects of the universe, including land, sentient beings and everything else. Nichiren Daishōnin identifies the character *myō* with 'the moon, the sun, and stars'. The extent of the power of the Mystic Law goes beyond the reaches of the Galaxy.

SHIMURA: I think I can now understand why Sessen Dōji sought the Buddhist teachings without regard for his own life.

IKEDA: This story teaches the importance of discipline which was espoused at the introductory stages of Buddhism. Moreover, *Nam-myōhō-renge-kyō*, as Nichiren Daishōnin elucidated, is the true entity of life permeating all phenomena in the universe and the ultimate principle underlying all the Buddhist teachings.

KIGUCHI: I assume that the theoretical understanding of Buddhism has its limitations, but an understanding of its all-encompassing concepts and universal ideas makes me feel even more awed by its profundity.

SHIMURA: By the way, there is a weekly magazine called *Nature* published in the United Kingdom. The editor of this magazine came to Japan to collect data for a special feature on Japanese scientific technology.

KIGUCHI: Yes, and Charles Darwin, who is famous for his theory of evolution, was involved in the publication of the first issue of the magazine in 1869. It is said that to have an article printed in this publication is a direct route to winning the Nobel Prize.

IKEDA: During his press interview, the editor said something to the effect that the field of quantum mechanics is becoming more important and that it may become *the* leading science in the future.

SHIMURA: He spoke about the developments in scientific research, saying that it is possible that an exciting discovery related to the fundamental nature of the physical world will be made based on progress in the field of quantum mechanics.

KIGUCHI: Dr. S. Chandrasekhar of the University of Chicago was awarded the Nobel Prize for his research concerning the prediction of the existence of white dwarfs, applying quantum mechanics to the field of astronomy. Quantum mechanics can be considered a science which is very close to philosophy in many ways. Therefore, it is conceivable that as this science advances further, human values will change accordingly.

NOTES: CHAPTER IX

1 Until 1873, Japan used the lunar calendar that it adopted from China. In the lunar calendar system, one year has about three hundred and fifty-four days; therefore, it deviates by about eleven days from the solar calendar. To compensate for this discrepancy, the Chinese invented the lunar-solar calendar by introducing seven intercalary months within a nineteen-year span. In this book, the lunar calendar refers to the lunar-solar calendar.
2 A Buddhist festival in which offerings are made to the three treasures for the benefit of the deceased, usually held on the fifteenth day of the seventh month (presently July; August in some areas). Also called *Urabon* or *Obon* (Skt *ullambana*). In Japan this custom merged with the belief that the spirits of the deceased are thought to return to the realms of their descendants. During this festival various folk dances (*bon odori*) are performed. During the *Bon* and New Year seasons, family members, no matter how far apart they may be, try to gather in one place for a reunion.
3 The mountains and seas which constitute the world, according to the ancient Indian view. The nine mountains are Mt. Sumeru at the centre of the world and eight concentric mountain ranges which surround it. These mountain ranges are separated by eight seas.
4 A chronicle of the Wei dynasty (220-265). This includes a section referring to 'Wa people' (the Japanese), which is one of the earliest references to Japan among Chinese historical records.
5 The present Japanese word for calendar is *koyomi*.
6 *Abidatsuma kusha ron* (Skt *Abhidharma-kosha-shāstra*).
7 10 November 1984 issue.

X

The Potential for Life and the Concept of Kū

The discovery of harmony and order existing within the universe, the cradle of life, as expounded by Buddhist doctrine.

Changes of the Seasons Lend Rhythm to Daily Life

SHIMURA: Dr. Kiguchi, I understand that, in ancient times, the first day of spring would occasionally fall before New Year's Day.

KIGUCHI: Yes. In the past the Japanese used the lunar calendar, in which a month was determined by the phases of the Moon, so it occasionally happened that spring arrived before New Year's Day.

IKEDA: I recall reading a *tanka* in the *Kokin wakashū*[1] (*Collection of Japanese Poems from Ancient and Modern Times*) which goes something like:

> Before the turn of the year
> spring has come.
> What should we call it,
> last year
> or this year?

SHIMURA: The first day of spring, which was traditionally regarded as the beginning of the year, came before New Year's Day on the calendar, so the author wonders what he should call the period in between.

IKEDA: This *tanka* expresses the bewilderment that can result from a time-lag in the calendar, the means people rely on to regulate their daily lives. The poem also has a touch of sarcasm.

KIGUCHI: It also shows how the lives of our ancestors were intimately associated with the rhythm of nature.

IKEDA: They were sensitive to seasonal changes and regarded the movements of the heavenly bodies as valuable signposts which enable them to make the necessary adjustments in their daily life patterns.

SHIMURA: Changes of the seasons are not as acutely felt today. We can now eat watermelons and strawberries even in winter, and tangerines and apples in summer, although they may not always taste as good.

IKEDA: The Gregorian calendar, or the solar calendar which we currently use, was adopted in the Meiji era (1868-1912) when the Japanese were attempting to become a part of the modern, industrialized world.

KIGUCHI: They determined that the day after the second day of the twelfth month of the fifth year of Meiji (1872) in the lunar calendar, would be January the first of the sixth year of Meiji (1873), and they went according to the Gregorian calendar from that day on.

IKEDA: According to the Oriental calendar, the first day of spring was New Year's Day. With the adoption of the solar calendar, however, New Year's Day arrived around the time of the winter solstice. But, because of a long tradition, Japanese people still tend to identify New Year's Day with spring.

SHIMURA: Under the old calendar system, the word 'new spring' was appropriate to the New Year season.

IKEDA: So, even after the calendar system was changed, people continued to use words such as 'new spring', 'happy spring' and 'the coming of spring' in their greetings on New Year's cards.

KIGUCHI: 'Spring' may be not the appropriate word for New Year in Japan, because it actually comes in the middle of winter, but then, countries throughout the world don't all greet the New Year during the same season, because the Earth revolves both around the Sun and on its own axis.

'Buddhism in Outer Space'

SHIMURA: Dr. Gerald P. Carr, a former American astronaut and captain of *Skylab 4*, paid you a courtesy call, Mr. Ikeda.

KIGUCHI: I had the honour of being present, and I must say that the talk between Mr. Ikeda and Dr. Carr was extremely interesting.

SHIMURA: I also had the pleasure of listening to your conversation, and had the same impression. Dr. Carr has spiritually and physically experienced the universe, the cradle of life. The discussion that he had with Mr. Ikeda might very well have been included as part of this dialogue on 'Buddhism and the Cosmos'.

IKEDA: It was an extremely important discussion, and quite enjoyable for me as well. Come to think of it, only a handful of people out of more than four billion inhabiting the Earth have actually experienced space flight – only around two hundred in the last twenty-odd years, I would imagine.

SHIMURA: That's the approximate number of American and Russian astronauts who have flown in space.

IKEDA: Dr. Carr is a humble man, and yet he is one of the pioneers in space research. I will long remember having had the opportunity to discuss the future of mankind and civilization with a person like him from a global standpoint.

KIGUCHI: I felt exactly the same way. For a man like myself, constantly involved in theoretical research, that discussion as a whole renewed my awareness, since I was listening to a person who had actually been in outer space.

SHIMURA: I was really surprised when, after Mr. Ikeda explained the Buddhist view of the relationship between the universe and human beings, Dr. Carr's instant reply was: 'That must mean that, when I was in space, I was learning about Buddhism.' He also said, 'Without a doubt, "harmony and order" exist in the universe,' and 'It certainly did *not* seem as if a god were pulling the strings and making things happen.' This was a very impressive remark.

KIGUCHI: I intuitively felt that what Dr. Carr experienced in space was connected with Buddhism.

IKEDA: He is well-known as having been the commander of the *Skylab* spacecraft. If I remember correctly, the spaceship docked with the *Skylab* space station and circled the Earth for eighty-four days, and Dr. Carr walked in space for a total of fifteen hours. He told me that the first thing he recognized when he exited the ship and saw the Earth was the boot-shaped country of Italy.

SHIMURA: When a person leaves a spaceship which is travelling at an extremely high speed, why doesn't he tumble out into space? Wouldn't he be left behind?

KIGUCHI: The spacecraft is not accelerating while it is in orbit, and so both the vehicle and the individual are moving at the same speed. This same principle applies when you drop an object from the mast of a ship moving at uniform velocity; the object falls directly under the mast.

IKEDA: Dr. Carr was kind enough to bring a NASA documentary film about the *Skylab* project. I thoroughly enjoyed seeing it the day after our meeting.

SHIMURA: That film, I would imagine, is a rare document which cannot be shown without NASA's permission.

IKEDA: There was a scene of Dr. Carr in a space suit engaged in activities outside the spacecraft, and his 'umbilical' cord and white tether line were visible against the background of space. On his left, the Earth appeared immense, wrapped in a gleaming pearl-grey atmosphere. On his right, space appeared as dark as black lacquer. I was overwhelmed at seeing the Earth set against the infinite universe in one shot.

SHIMURA: I understand that it takes ninety minutes for a spaceship to orbit the Earth.

IKEDA: Yes. This means that you witness fifteen sunrises and fifteen sunsets in the course of a twenty-four-hour day, so it affects one's orientation in a way that's beyond one's imagination while on Earth. All of a sudden you enter a new world, one that is beyond normal human conceptions.
   It reminds me of some of the incredible events described in the *Lotus Sutra*, such as the Buddha purifying and transforming the *sahā* world in which we live and the two other groups of innumerable worlds into Buddha lands and then levitating the entire assembly on Eagle Peak into mid-air.

SHIMURA: I hear that the farther away from the atmosphere a spaceship travels, the more the sunrise appears like a ring of shimmering light.

IKEDA: Dr. Carr spoke about that. It must be a spectacular sight.

SHIMURA: But, once the spacecraft moves behind the Earth, everything suddenly becomes pitch-black.

KIGUCHI: That's because there is no atmosphere in space. On Earth, sunlight is refracted and scattered by the atmospheric envelope, but in space, light travels in a straight line. The sunrise seen from space must be far more dramatic than it is anywhere on Earth.

SHIMURA: Sunsets seen from space are reported to be nothing less than spectacular. I remember Dr. Carr leaning forward enthusiastically, using gestures to add emphasis to his descriptions. He could see a thin layer of atmosphere embracing the curving horizon of the Earth and the Sun descending to the other side. At that very moment, many bright colours suddenly burst forth, and, according to Dr. Carr, the scene was breathtakingly beautiful.

KIGUCHI: That's because the atmosphere, which envelops the Earth, acts like a huge prism, which breaks the light down into red, yellow, green and blue.

SHIMURA: Dr. Carr offered the same explanation, saying he was completely overwhelmed by the magnificent spectrum of colours.

IKEDA: People have a tendency to try to find an explanation for such wonders. And the more we learn about the greatness of the universe, the more acutely we sense how petty and meaningless are the many conflicts and wars taking place on Earth. I hope that the space age will mark the dawn of an era in which all countries and peoples of the world will renounce war forever. We must do everything we can to achieve this. The space age will compel us to consider everything from the standpoint of the human race as a whole.

SHIMURA: Dr. Carr told us that every ten seconds, he had taken pictures of the atmosphere as it changed colour. When the sun sets completely, it suddenly becomes pitch-dark, and at exactly the same moment, innumerable stars appear. I remember Dr. Carr's remark that, even if we gathered all the jewels on Earth and spread them out on black velvet, they could not possibly outshine the beauty of the stars.

KIGUCHI: Dr. Carr agreed that there is a strong possibility that extraterrestrial life exists.

IKEDA: He went as far as to say that superior, intelligent life might exist. He said jokingly that they might be observing us and wondering what these creatures on Earth are up to.

In any event, it must have been an incredible experience for him. He must have perceived more of the mystery of life in a moment, by experiencing first-hand the wonders of the universe, than most people do during the course of their entire lives. And while he may not have been able to concretely theorize and systematize his own concepts about human life or the universe, he seems to have come to the conclusion that he is personally inclined toward Buddhism.

KIGUCHI: I agree. Dr. Carr's experience is precious and meaningful, and I believe he was extremely pleased to have met in Japan a person to whom he was able to relate his experience in a straightforward manner.

SHIMURA: I heard later that on his way back to his hotel after the discussion, Dr. Carr commented, 'When viewed from an orbiting spacecraft, the size and shape of both the Big Dipper and the Southern Cross, which are used for navigation on Earth, appear different, making it difficult to get a fix on your position. In the same way, modern society has lost track of its fixed points of reference which would enable it to solve problems and secure peace.'

I also heard that, back in his hotel, he continued to talk with his wife about the interview until late at night.

KIGUCHI: It was a memorable evening for me, as well.

SHIMURA: Dr. Carr said, 'I met many people during my visit to Japan at lectures and on other occasions, but the most memorable event was meeting Mr. Ikeda; we had wonderful time together.'

KIGUCHI: The stars seen from the spaceship must be really beautiful. They must emit thousands of times, or even tens of thousands of times, the radiance we observe from the Earth.

IKEDA: They must, because there is no atmosphere. There are eighty-eight internationally recognized constellations, and eighty-five of them can be seen from Japan.

KIGUCHI: Eighty-four can be seen even from Tokyo, and some are constellations not clearly visible in other countries.

SHIMURA: I imagine that the atmosphere is clearest in winter, and the stars also look the brightest. Winter must be the best season for amateur astronomers to observe the constellations.

IKEDA: My third son is a great astronomy enthusiast. When he was attending elementary school and junior high school, he used to sneak out into our garden at midnight in February, all bundled up in a blanket, to peek through his telescope. This was much more important to him than preparing for school examinations. At all events, the best time for viewing the stars seems to be cold winter nights.

The stars appeal strongly to the human heart. I remember that Einstein once expressed great admiration for Schubert's work entitled *The Stars*. Since Schubert was a gifted composer, when he looked up at the heavens, his heart must have been filled with an overwhelming desire to express his poetic sentiments.

Speaking of the stars in winter, there is the Big Dipper which connotes seasonal changes, referred to in *haiku* poetry as the 'winter Dipper' and the 'midwinter Dipper'. There is also Orion, and the 'midwinter Pleiades' which appear in the Japanese classic, *The Pillow Book*.[2] Moreover, there are many magnificent stars like Sirius, the brightest fixed star in the heavens.

KIGUCHI: In Japan we are indeed fortunate that we can see almost all the constellations if we continue to observe the sky throughout the year.

SHIMURA: We should instil in our children an appreciation for the workings of nature.

IKEDA: I believe it is an important part of education to teach the young how to appreciate the stars, the sea, mountains and nature at every opportunity. When my third son was in high school, he asked to visit Chichi-jima and Haha-jima of the Bonin Islands during his summer vacation with a friend, so that he could watch the stars. At that time, it took more than thirty hours to reach these islands by ship, and my wife was somewhat worried. But I convinced her that this would be a much better experience for a fast-growing boy than to just let him roam the crowded streets of Tokyo, and so we permitted him to go. As it turned out, the splendour of the night sky, such as can never be observed in large cities, made a very strong impression on him.

SHIMURA: By the way, I understand that, immediately after Japan opened its doors to the rest of the world in the Meiji period, many of the world's prominent astronomers came to visit Japan.

IKEDA: I have heard that John Evershed (1864-1956), the noted British solar physicist, and Percival Lowell (1855-1916), an American astronomer renowned for his observations of Mars, visited Japan.

KIGUCHI: These two scientists are great names in the history of astronomy.

SHIMURA: I also checked, and found that Herbert H. Turner (1861-1930) of the United Kingdom and John M. Schaeberle (1853-1924) of the United States came to Japan as well. They both wrote about their impressions of Japan. I believe that the accounts written by Lowell have been translated into Japanese.

IKEDA: My understanding is that Lowell came to Japan on business, but I read somewhere that he made a total of four trips, and each time he brought his telescope with him.

SHIMURA: The travelogue he published after touring the various places along the Japan Sea coast is a unique account of Japan as observed by a foreigner. It is amazing that a number of astronomers visited Japan at a time when it was still considered an 'uncivilized' country. We can see that the attraction of the stars was great.

KIGUCHI: It has long been said that people who observe and research the cosmos are all romantics, and that there are no disreputable individuals among them.

IKEDA: There was an author in ancient Greece who wrote a story in which the hero is so fascinated with the world of the stars that he ends up landing on the Moon. This writer is also thought to have been an extremely good-natured man.

SHIMURA: You are referring to Lucian, who lived in the second century AD. This novel is about a man whose ship gets caught in the funnel of a large tornado, is then sucked into the heavens and eventually lands on the Moon.

KIGUCHI: It sounds somewhat preposterous, but perhaps we can call it a classical version of the science-fiction novel.

SHIMURA: But Lucian actually thought that it was quite possible. This is evident from the title he gave to the work, which is *Vera historia*, meaning 'A True Story'.

Kant's Interest in the Lotus Sutra

SHIMURA: I recall that Kant also wrote an essay on Japan.

IKEDA: He wrote about his affinity for Japan, based on a travel sketch by a German merchant who visited Japan during its period of isolation from the rest of the world toward the end of the Edo era (1600-1867).

SHIMURA: But Kant's longing was of a different nature than that of Marco Polo or Columbus, both of whom sought Cipango,[3] the country thought to be abundant in gold.

IKEDA: The interesting point is that Kant stated, using a very simple expression, that the leading scripture of Japanese Buddhism is the *Buch der Blumen* (*Flower Scripture*). He used this expression to indicate the Sutra of the Flower of the Law, or *Lotus Sutra*. I cannot help concluding that he took a strong interest in this sutra.

SHIMURA: The Guimet Museum in France is well-known for its collection of works of art from the Orient. And I have heard that the orientalists there made a general study of Japanese Buddhism, and that the fruit of their research was a publication entitled *The Doctrine of Nichiren*.

IKEDA: Of course, it cannot have completely grasped the heart of the Buddhism of Nichiren Daishōnin, but it still indicates the extent to which research was done. It was supported by the French Ministry of Education, and was published in 1953, after its final compilation by a scholar of oriental studies named Gaston Renondeau.

I understand that there is a passage in that book which reads: 'He [Nichiren Daishōnin] set out to restore Buddhism to its orthodoxy and to make Japan the centre from which the truth would eventually spread to the rest of the world ...' This is a recognition of the universality of Buddhism.

Another feature we should note is that this book contains the translation of several selections of the Daishōnin's writings, including the *Risshō ankoku ron* (*On Securing the Peace of the Land through the Propagation of True Buddhism*), *Kaimoku shō* (*The Opening of the Eyes*), *Kanjin no honzon shō* (*The True Object of Worship to Observe One's Own Life*) and *Hokke shuyō shō* (*The Essentials of the Lotus Sutra*). Through this book, whatever the interpretations may be, many French people came into contact with the Daishōnin's writings for the first time.

SHIMURA: I hope we will have the opportunity to have this valuable document translated into Japanese.

IKEDA: Come to think of it, Dr. Toynbee, in his book, *A Study of History*, devotes quite a few pages to Nichiren Daishōnin and the religious climate of the Kamakura period (1185-1333). This, I am sure, was one of the reasons for his desire to speak with me.

How Will the Earth End?

SHIMURA: Incidentally, a peculiar star, which might be a black hole, was discovered by Japanese scientists in November 1983.

KIGUCHI: It was sighted by an astronomical satellite, *Tenma*[4] (*Pegasus*), carrying an X-ray telescope.

IKEDA: Black holes have attracted much interest. When I was in Hong Kong in December 1983, there was a planetarium right across the street from my hotel. I heard the astronomers there were actively involved in black hole research. Unfortunately, my schedule was such that I could not take advantage of the opportunity to visit the planetarium.

KIGUCHI: Black holes absorb all light and matter within their surroundings, so they are extremely difficult to locate.

IKEDA: I have heard that only one has been discovered so far, within the constellation Cygnus, by an American observatory.

KIGUCHI: So far, Cygnus X-1 is the only object which most astronomers recognize as a genuine black hole. But, in November 1982, what is suspected to be a black hole was discovered within the Large Magellanic Cloud. If the discovery by the *Tenma* turns out to be a black hole, it may be the third such object to be found.

IKEDA: The decisive clue which led to the latest discovery, I understand, was that the luminous gas from a neighbouring star was seen flowing toward a whirlpool within the nebula, indicating the possible existence of a black hole.

KIGUCHI: When a star is located in the immediate

neighbourhood of a black hole – in other words, when the black hole is one component of a binary – the star will be visibly affected by the black hole, and the latter is, therefore, relatively easy to locate.

SHIMURA: The black hole, the aftermath of the death of a star, is considered to be one of the greatest mysteries. Will our Sun experience the same fate at the end of its life?

KIGUCHI: Given the size of the Sun, whether or not it will become a black hole is a complicated question. If it were larger, then it would definitely become one.

SHIMURA: The death of a star larger than the Sun seems far beyond our imagination.

KIGUCHI: Even in the case of our Sun, when it enters into its final stages of existence, the Earth will share the same fate, becoming an inferno, a scene often described in science-fiction novels.

IKEDA: There's a novel out by an American author depicting the end of the world in detail.

SHIMURA: The title is *The Last Day*, by Richard Matheson.

IKEDA: It describes how our civilized world, which mankind has laboriously developed over a period of some five thousand years, collapses in a matter of only fifty hours or so.

SHIMURA: Dr. Kiguchi, how does the field of astronomy view the death of the planet Earth?

KIGUCHI: It's thought that first of all, the Sun will start to expand and then swallow Mercury and Venus, which are the two closest planets. On the surface of the Sun, great explosions will occur everywhere. The Earth will be completely scorched. The total destruction of the Sun will take place approximately 5.5 billion years from now.

IKEDA: Have any of the fixed stars, like the Sun, in our galactic system ever been observed blowing up in a huge explosion and then vanishing into space?

KIGUCHI: Yes, many. Among them, the Crab Nebula is an example.

SHIMURA: Why does a star expand so greatly in its final stages?

KIGUCHI: The Sun, for instance, shines a golden yellow due to the fusion reaction of hydrogen – in other words, the conversion of hydrogen into helium. It is, so to speak, a huge atomic power plant, built by nature.

SHIMURA: So it radiates incredible amounts of light and heat, using hydrogen as its fuel.

IKEDA: When man succeeded in tapping the power of the atom, nuclear arms came into being. Human beings must now determine whether this power of the universe is to be used for good or evil. I once wrote in a magazine article: 'The purity of human hearts can lead to eternal peace, but the baseness of human nature can create an eternal hell of conflict and antagonism.' This is a crucial matter which we should seriously ponder in this nuclear age of ours.

KIGUCHI: The Sun's hydrogen is destined to be consumed because the supply is limited. Before burning out completely, the Sun will show signs of deterioration. As hydrogen is used up in the process, the helium resultant, which is the waste product of fusion, will accumulate in the Sun's core.

IKEDA: Is it something like the cinders of burnt coals?

KIGUCHI: Sort of. The Sun's core becomes bottled up with these waste products. Then, as the outer section continues to burn and expand, the core begins to contract. Thus, as the exterior of the Sun expands, the Sun's surface temperature decreases. The emitted light, which was yellow, now becomes red.

IKEDA: Is that why such stars are called red giants?

KIGUCHI: Exactly. Observations have revealed that the radius of one such star, Antares of the constellation Scorpio, has expanded to one hundred and seventy-five times that of the Sun – in other words, to more than one hundred and twenty million kilometres.

IKEDA: That seems beyond anyone's ability to comprehend. Then, when the Sun enters its final stages, how far will its radius expand?

KIGUCHI: According to my calculations, it would probably be over seventy million kilometres, which is approximately one hundred times its present radius. And it could possibly be double this figure.

SHIMURA: If the Sun grows that large, I would imagine that its outer surface would come closer and closer to the Earth, exerting a devastating effect on it.

KIGUCHI: Yes. Since the distance between the Earth and the Sun is only 149.6 million kilometres, the effect would be catastrophic. The oceans would begin to boil fiercely, and the atmosphere would dissipate completely.

IKEDA: Isn't the cubic volume of the Sun approximately 1.3 million times that of the Earth?

KIGUCHI: That's correct.

IKEDA: The spectacle of such a huge object exploding is simply beyond the imagination. There is a passage in the *Lotus Sutra* which reads: 'The world is in no way firm or secure, but it is like water-bubbles, like a will-o'-the-wisp!' Neither the Sun nor the Earth is an exception to this rule.

The Death of the Galaxy and Jōsei Toda's Insight

IKEDA: Immediately after the war, the late Jōsei Toda, second president of the Sōka Gakkai, used to talk about his ideas concerning the disintegration of the galactic universe. I remember being simply stunned by the scale of it.

His conceptualization was as follows: 'First, the solar system might cease all of its movements. Or it might speed up toward another star. The same thing could happen to other fixed stars as well: Their movements might be interrupted, or they might accelerate. Smoke and gas would envelop the entire universe, and there would be lightning everywhere, generating intense heat which would surpass all imagination. The cosmos would thus be plunged into a state of utter confusion. Molecules would undergo a cycle of decomposition and recomposition. It would be a magnificent, or should I say awesome, spectacle which one can only attempt to imagine, for not a single person would survive to witness it.'

SHIMURA: This statement is recorded in the collected essays of Mr. Toda. I was astonished that this article, which describes a phenomenon as awesome as the death of the galactic universe, is entitled 'A Small Discovery'.

KIGUCHI: Mr. Toda seems to have possessed great insight. When did he make those remarks?

SHIMURA: The essay is dated 1 August 1946.

KIGUCHI: Then I must say that he had tremendous vision.

IKEDA: Mr. Toda was a great teacher of mathematics. Moreover, he had profound insight and knowledge in many fields of learning.

SHIMURA: I heard that, because the explosion of stars occurs abruptly in a corner of the night sky, until recently people mistook them for the birth of new stars.

KIGUCHI: That's right. They're even named *supernovae*, or new super stars, because following the explosion, they emit such intense light that it can be observed with the naked eye.

SHIMURA: It seems ironic that the death of a star is mistaken for the birth of a new one.

KIGUCHI: Their intense light quickly dissipates, and eventually they merge into the universe. But then, were it not for this phenomenon, new stars could not come into being.

The Buddhist Concept of Kū

IKEDA: We have mentioned this before, but like all other phenomena, stars, too, are born and die according to the Buddhist principle that everything invariably undergoes the cycle of birth and death. After an enormous length of time, the residue which remains after the death of a star coalesces to create a new star similar to the Sun. In other words, the matter remaining after the death of a star is the same matter which comprises a new star.

KIGUCHI: Seen through a radio-telescope or infra-red telescope, space is filled with innumerable, extremely minute particles. Some scientists term them 'ultra particulates'. These substances gather to form a star.

SHIMURA: They did not exist when the cosmos first came into being, but they appeared as if from nothingness. When we see the word *kū*, we usually interpret it as literally meaning 'nothingness'. In reality, however, according to the Buddhist teachings, *kū* is far from nothingness. On the contrary, it is something like the matrix which gives birth to everything.

IKEDA: Even if all matter and all phenomena in the universe should perish, this would not mean that everything has ceased to exist. It would mean, instead, that the 'nucleus', or the 'self', which constituted such matter and phenomena, has fused into the state of *kū*.

KIGUCHI: Some characteristics of *kū* can be observed in the way radio waves are received by our television sets and radios, although we watch or listen to them every day without even thinking about it.

SHIMURA: Radio waves are a mysterious phenomenon. Even though we cannot see them, this room is filled with radio waves transmitted from all over the world.

IKEDA: Long waves, medium waves, short waves, ultra-short waves – in fact, all kinds of radio waves are criss-crossing one another at this very moment. Even more incredibly, once you adjust the frequency of your receiver to a given wavelength, and the information in the signal is demodulated, a sound or picture is produced.

KIGUCHI: Radio waves are propagated through space. At first glance, space appears to be empty and void of activity, but in reality, space is bustling with activity. Radio and television communications utilize that property of space which transforms electrical and magnetic energy. Understanding these natural phenomena makes it easier for me to approach the concept of *kū* as expounded in Buddhism.

IKEDA: Radio waves transmitted from Cape Kennedy reach the *Apollo* spacecraft and beyond, continuing out into space. Similarly, the prayers we offer to the Gohonzon, the embodiment of the ultimate Law of the universe, will, like radio waves, penetrate the entire universe. In the light of this principle, the doctrine of *ichinen sanzen*, or three thousand realms in a single life-moment, becomes clear.

SHIMURA: During the period between December 1982 and

January 1983, scientists at the Nobeyama Radio Observatory in Nagano Prefecture discovered that a star is being born within the constellation Orion.

KIGUCHI: It's taking place within an astronomical time-frame. They discovered that gas has started to flow into where there had thus far been nothing. They also found that a rotating disc is being formed. This indicates that a star is in the making.

IKEDA: Invisible particles of dust come into being within the cosmos, and combine to form a star, which eventually becomes extinct. We can understand the doctrine of the three truths even more clearly when we view it in terms of the dynamics taking place within the infinite expanse of the universe.

KIGUCHI: Even when it appears that everything has ceased to exist, space still retains the potential to generate matter wherever the cosmos evolves sufficiently to offer the right conditions. Research in quantum mechanics has revealed that space is not nothingness. It is in this connection that the disintegration of protons has recently attracted attention.

IKEDA: Certainly, nothing is born from nothing. We must recognize that only the Buddhist concept of *kū* (potential or latent state) can shed light on this issue.

Space, which had been considered a vacuum in which nothing exists, has turned out to be, in reality, a life-containing entity capable of generating matter. This discovery is one of the greatest achievements of science during the twentieth century.

KIGUCHI: This was theoretically demonstrated by the Nobel Prize-winning British physicist, Paul A.M. Dirac. It is one of the greatest theories proposed in this century, as Mr. Ikeda has just pointed out.

SHIMURA: The fact that substance is created from space which is seemingly void, and that this space yields galaxies and gigantic stars, suggests the Buddhist concept of *kū*, if I understand it correctly.

IKEDA: The Buddhist concept of *kū* in no way indicates emptiness or nothingness. To explain it in simple terms, Buddhism teaches that everything is initially in the state of *kū* or latency. All things come into existence out of the state of *ku*, and then perish to merge with the state of *kū*. Therefore, to use a rather complicated expression, *kū* may be defined as the 'reality which transcends the concepts of existence and non-existence'.

KIGUCHI: That reminds me of the field theory postulated by Einstein. The theory states that, in any part of the universe, there is invariably a physical law at work to generate matter. Einstein termed the location where this law operates a 'field'.

SHIMURA: I feel that this theory grasps, though on a different level, at least one aspect of the concept of *kū* which governs the universe.

KIGUCHI: I think that the field theory will provide a foothold in our attempts to resolve the mysteries of the universe. Mr. Ikeda, could you please explain in more detail the 'reality transcending the concepts of existence and non-existence'?

Kū: *Life-oriented Space Containing Infinite Creativity*

IKEDA: Briefly speaking, *kū* is space which has the potential to manifest matter or energy. By coming in contact with the appropriate external cause – in other words, under the right conditions – it becomes activated, goes into operation and begins to create. In consequence, *kū* may be called life-oriented, apparently vacant space which contains infinite potential and boundless creativity.

SHIMURA: I presume this concept of *kū* is unique to Buddhism.

IKEDA: I think so. It is not found in Western religions, philosophy or thought.
With the progress of science, various new truths have been discovered, some of which cannot be explained according to

Western logic. So it becomes necessary to turn to the Buddhist concept of *kū*.

SHIMURA: That is why, after the war, prominent scientists such as Heisenberg, Bohr and Einstein began to seek answers in Buddhism.

When we die, our bodies cease to exist in the phenomenal world. Can this state be called *kū* rather than nothingness?

IKEDA: Yes. *Kū* can be seen as an abstruse and profound Buddhist philosophy which embraces the mysterious existence of life. The concept of *kū* is underlain by the vast amount of sutras, categorically referred to as the eighty thousand teachings. It is one of the three truths, namely, *kū* (*kūtai*) or latency, *ke* (*ketai*) or temporary existence, and *chū* (*chūtai*) or the Middle Way.

According to one explanation in Buddhism, human beings and all other things exist in the phenomenal world in their respective forms and shapes because the necessary components are temporarily combined in their individual forms through the function of *innen*, or internal and external causes.

KIGUCHI: *Innen* is also translated as 'dependent origination'.

IKEDA: You can call it that. Internal cause is the principal agency which produces an effect while external cause serves to manifest that effect. Thus all things exist in the form of a temporary union as the result of the conjunction of internal and external causes. This is called *ketai*.

KIGUCHI: What does *tai*, or 'truth' of the three truths, mean?

IKEDA: It means 'true and clear' or 'perception of the eternally unchangeable truth'. In other words, it means to clearly penetrate, based on the eternal Law, the truth of the universe on a macrocosmic scale and the truth of life on a microcosmic scale.

SHIMURA: Then, what is the significance of *kūtai*?

IKEDA: In short, it indicates the properties or qualities of all things. Everything that has form and shape possesses characteristics of its own. For instance, consider an elementary particle. No matter how small it may be, it has its own unique nature.

Ketai, or the temporary union formed through the conjunction of internal and external causes and having its own unique characteristics (*kūtai*), does not permanently remain in this state but is destined to eventually perish. However, even after it loses its form and shape, *kūtai*, or its essential properties and qualities, will continue to exist as attributes unique to that existence.

KIGUCHI: Then, are you saying that even after our bodies cease to exist, *kūtai*, or the workings of our individual nature or life-tendency, endure eternally?

IKEDA: According to the teachings, yes. *Kūtai* is dualistic. Either it finds itself within a living being (*ketai*), or it fuses itself with the unvierse when *ketai* ceases to exist. While manifested within a living being, *kūtai* becomes the driving force of progress, vitality and infinite creativity.

SHIMURA: An entity of life in the state of death can neither be perceived nor grasped.

IKEDA: That is true. *Kūtai*, however, contains an unchangeable 'nucleus' of life itself. This 'nucleus', or that which Buddhism calls the 'self', continues to exist for all eternity.

KIGUCHI: As you said, 'Even after we have died physically, the essence of life itself, namely the "self", continues to exist.'

SHIMURA: I now feel that I can comprehend the 'reality transcending the concepts of existence and non-existence'. What, then, is the meaning of *chūtai*?

IKEDA: *Chūtai* can be thought of as the permanently

unchangeable 'nucleus' of life, which I have just mentioned. Both *kūtai* and *ketai* are essentially fused in the 'self' as *chūtai*.

What is the ultimate entity which constitutes and activates the 'self'? Nichiren Daishōnin gives a clear-cut and concrete answer: 'It is the mystic entity of the Middle Way that is the reality of all things.'

SHIMURA: By the way, we often use the word 'nucleus', but what is the nature of an atomic nucleus? It's invisible to the naked eye. How big is it?

KIGUCHI: It is about one ten-trillionth of a centimetre. As for weight, in the case of protons for instance, a trillion multiplied by a trillion protons would weigh only 1.6 grams.

IKEDA: It's amazing that our bodies are the aggregates of such tiny particles.

KIGUCHI: We now know, however, that even these invisible atomic nuclei undergo the cycle of formation and decay. Protons were thought to be indestructible, but they, too, are now known to disintegrate. Modern physics is now looking into the unchangeable law that governs the formation and extinction of atomic nuclei.

IKEDA: In the light of such laws of physics, we can clearly understand *chūtai* which exists as the unchangeable 'nucleus' of life itself.

SHIMURA: There is also the concept of karma in Buddhism.

IKEDA: It is part of the concept of internal and external causes. *Maka shikan* (*Great Concentration and Insight*) states, 'That which brings about a latent effect is called an internal cause, which may also be termed karma.' The 'latent effect' is the one produced and stored in the depths of one's life until, triggered by some external cause, it becomes manifest. As the 'self' undergoes the endless cycle of birth and death, it establishes a certain tendency.

SHIMURA: Is it something like one's habits or idiosyncracies?

IKEDA: Put simply, yes. Karma is classified in many ways. It is divided into two types: one which one has formed in his previous existence, and the other which he is forming in this lifetime. Both types of karma serve as causes that will produce effects in one's future existence. The state in which karma is dormant, being carried over to the next existence, is *kūtai*.

SHIMURA: If that is the case, can we say that life in the state of *ketai* and death which is the state of *kūtai* are two aspects which the 'self', or *chūtai*, manifests?

IKEDA: In a way, we can. Moreover, the ultimate doctrine of Buddhism teaches that the three truths are inseparable. If you consider a human being from the standpoint of the oneness of body and mind, his mental or spiritual aspect is *kūtai*, his physical aspect is *ketai*, and his life itself is *chūtai*, but the three are inseparably unified to form the living being.

Thus, each of the three states – life as *ketai*, death as *kūtai*, and the permanently unchangeable entity of life as *chūtai* – also embodies the three truths. The state in which all these sets of the three truths are harmonized with one another is referred to as the unification of the three truths (*en'yū no santai*). This is one of the most profound of all the teachings of Buddhism.

Nichiren Daishōnin states, 'What is the reality expressed by the unification of the three truths? It is none other than *Nam-myōhō-renge-kyō*.'

Obtaining 'the Supreme Cluster of Jewels'

IKEDA: Thus Nichiren Daishōnin teaches that the true entity of life, the universe and everything within it is *Nam-myōhō-renge-kyō*. This is the Law which enables all things and all phenomena to maintain perfect harmony and order. It is also the source of limitless energy for creation and revitalization. In a word, *Nam-myōhō-renge-kyō* is the Law which ensures that human beings and society continue to develop and progress.

The Daishōnin embodied the Law of *Nam-myōhō-renge-kyō*, the reality expressed by the unification of the three truths, in the form of a mandala. This is the object of worship in Nichiren Daishōnin's Buddhism. Therefore, as one continues to chant *daimoku*, or the invocation of *Nam-myōhō-renge-kyō*, to the object of worship, his personality is gradually purified and perfected. The wisdom and desire to develop one's own personality are also oriented toward ensuring the happiness of other people and the peace of society and the world.

Everyone wants to become happy and actively seeks fulfilment. But our earthly desires, karma and suffering prevent us from attaining it. We may have occasional feelings of satisfaction, but even this does not last long.

The immense power of chanting *daimoku*, however, will bring about a complete turnaround. With this we can obtain the wisdom to control our earthly desires and eventually become emancipated from the shackles of karma. Our personalities will become more and more polished as we squarely face our sufferings and overcome them. Not only will we be following the path to our own happiness, but we will also be acting in such a way as to contribute toward the greater welfare of others and society.

Nichiren Daishōnin, therefore, likens taking faith in the Law of *Nam-myōhō-renge-kyō* to obtaining 'the supreme cluster of jewels when one least expects it.' Through faith in this Law, any person can enter into a state of happiness which previously he could not even have imagined to exist. Moreover, the Daishōnin teaches us how to practise the Law in a perfectly reasonable way. *Nam-myōhō-renge-kyō* is truly the essence of Buddhism.

KIGUCHI: I have always thought that Buddhism, a profound philosophy, offers a clear path toward self-perfection.

IKEDA: Almost everyone is aiming toward a life of fulfilment, but during this process, one must inevitably suffer the results of the negative causes he has sown in past existences, that is, one's karma. Human existence is precarious. Change is the way of this world. Just because one is happy today is no guarantee that he will continue to be happy tomorrow. So, instead of seeking relative happiness, we should always strive to develop ourselves and to elevate our life-state. This determines whether or not we are truly living meaningful existences as human beings. Hence the need to have the spirit to seek a clear guideline, based on the indestructible Law of life.

NOTES: CHAPTER X

1 *Kokin wakashū*: a Japanese anthology consisting of some 1,100 poems, compiled by imperial order in the early tenth century.
2 *The Pillow Book* (Jap *Makura no sōshi*): prose by Sei Shōnagon, a tenth-century Japanese court-lady, consisting of short narratives, informal essays, impressions, etc.
3 Cipango: the name by which Japan was known to medieval Europeans. It first appeared in the *Book of Marco Polo*.
4 *Tenma*: an astronomical satellite launched on 20 February 1983 from Uchinoura, Kagoshima Prefecture, by the Institute of Space and Aeronautical Science (then part of Tokyo University).

XI

The Nuclear Threat and the Buddhist Philosophy of Peace

What is the true nature of nuclear power and its influence? And how should human beings cope with it?

The Vast Expanse of the Universe

SHIMURA: Recently we have witnessed the discovery of some two hundred thousand new stars and twenty thousand nebulae as a result of the infra-red astronomical satellite (IRAS) mission.

IKEDA: Human awareness has now extended to realms of the universe invisible to the naked eye.

KIGUCHI: These discoveries have resulted from observations made with a high-power, infra-red telescope. Individuals involved predict that by 1990 we will be launching satellites with more than a thousand times the performance capacity of those in use today.

IKEDA: Such discoveries may not have much immediate impact on people's daily lives, but they must surely bring about a change in people's value structure and way of thinking.

KIGUCHI: I believe that is so. In the past the emergence of new views of the cosmos has often been accompanied by major changes in society.

SHIMURA: The response to the discoveries of Renaissance scientists such as Galileo and Copernicus is fairly representative.

KIGUCHI: The discovery that the Earth revolves around the Sun heralded the dawn of modern science.

The Progress of the Telescope

SHIMURA: Dr. Kiguchi mentioned the subject of astronomical observation satellite. It seems we have quite a number of man-made satellites orbiting the Earth.

KIGUCHI: I don't know exactly how many, but I've heard it's about three thousand. Including those that are no longer operational, the figure must be close to ten thousand. For example, I'm sure you are familiar with television weather forecasts from the stationary satellite *Himawari* (*Sunflower*). Launched above the equator and facing east, it completes one orbit every twenty-four hours, together with the Earth's rotation. So-called reconaissance satellites can photograph an area of the Earth's surface as small as ninety thousand square metres. Then, if one wishes to see a certain part in greater detail, it can be magnified up to thirty times.

SHIMURA: So one could, for example, observe the figures of people strolling on Broadway?

KIGUCHI: Actually, you could obtain a close-up of someone waiting for a traffic signal to change and smoking a cigarette, say, at Times Square. That's how precise these satellite transmissions are. That's why they are now being used to search for resources under the Earth's surface using infra-red or X-rays.

IKEDA: If one observes from a high place with a clear field of vision, he must feel as though he can see everything. Traces of a 'phantom river', which appears in Egyptian legends, were discovered from photos taken by the space shuttle *Columbia*.

SHIMURA: That was an amazing discovery – traces of a river from several hundred thousand years ago actually remain under the desert west of the Nile!

KIGUCHI: By launching these satellites fitted with X-ray or infra-red telescopes, it has become evident that the universe has an expanse tens of thousands of times greater than what we are able to discern, observing it from the surface of the Earth.

SHIMURA: Dr. Carr reportedly sighted a forest fire in the mountains when his spaceship approached the airspace over Australia.

KIGUCHI: When he communicated with the Earth, the fire had not yet been detected because it was deep in the mountains. His sighting was greatly appreciated. By the way, I understand that the first Japanese to look through a telescope was Oda Nobunaga[1] (1534-1582).

IKEDA: It is on record that about four hundred years ago, in 1574, a European visitor presented him with an optical device that augmented vision considerably.

SHIMURA: That would have been several decades before Galileo discovered the moons of Jupiter, the craters and valleys of the Moon, sunspots and other phenomena. Of course, the spyglass Nobunaga received was not intended for observing heavenly bodies. It was simply a harbinger, so to speak, of the telescopes to come.

IKEDA: Well, I suppose it was an extremely precious object in those days – a suitable gift for the ruler of a foreign country. It is also on record that another one was presented to Tokugawa Ieyasu[2] (1543-1616).

SHIMURA: That brings to mind an amusing anecdote from the Edo period (1600-1867). One day, a certain feudal lord was gazing through a telescope from his castle tower. When he trained the telescope on the castle below, the figures of two men engaged in earnest conversation sprang into view, whereupon the lord unthinkingly put the telescope to his ear.

'Nuclear Force'

SHIMURA: From a time when such amusing incidents of that sort occurred, science advanced by giant steps, until at last, through the study of the stars, nuclear energy was discovered.

KIGUCHI: That's true. Forty-odd years have already passed since an atomic reactor was first tested in 1942, resulting in the control of atomic power.

IKEDA: The discovery of this power inherent in the cosmos has had a tremendous impact on people, whether they realize it or not. For most, however, the very words 'nuclear energy' carry an image of vast destructive power. At the same time, the complexity of this subject has confined thoughtful discussion of it almost exclusively to the scientific community. I wonder if this isn't the reason why relatively few people are motivated to actively come to grips with the problem of nuclear regulation.

SHIMURA: It seems so complex, somehow, as to be beyond our powers of comprehension. Dr. Kiguchi, would you please explain the concept of nuclear energy to us in simple terms?

KIGUCHI: Simply stated, all matter existing in the universe is composed of atoms. Moreover, each atom has a nucleus, around which a number of electrons 'orbit', so to speak.

SHIMURA: We can visualize this as similar to the model of the solar system, in which planets orbit around the Sun. It is amazing that something like the solar system should exist within anything as small as the atom.

KIGUCHI: Moreover, in terms of volume, the greater part of the atom consists of unoccupied space. One can say atoms are 'empty', in a sense. By analogy, if each of us were an atomic nucleus, the person next to us would be ten kilometres away.

SHIMURA: Then our own bodies which are composed of countless atoms are also virtually 'empty'. This is difficult to conceptualize.

KIGUCHI: In this light, I feel one can understand what in Buddhism is described as the five elements of the human body – earth, water, fire, wind and space.

SHIMURA: Yes, that would be a natural connection to make.

IKEDA: Our own bodies contain countless atomic nuclei and their orbiting electrons. The body is, so to speak, like a vast collection of tiny solar systems. Truly, one could call it a universe in miniature.

KIGUCHI: Moreover, the nucleus of the atom is in itself a universe, composed of elementary, or subatomic, particles such as protons and neutrons. These protons and neutrons are held together by a powerful force of attraction termed 'nuclear force'. It is one of the four major forces operating in the universe. I believe I may have mentioned this before, but all matter in the cosmos is governed by four basic physical forces: gravitational interaction, discovered by Newton; electro-magnetic interaction, described by Maxwell; the weak interaction, elucidated by Fermi; and the strong interaction, or nuclear force, theorized by Dr. Yukawa. Among these four, the most powerful is the nuclear force, which holds together the nucleus of the atom.

IKEDA: And it is this power that atomic bombs employ?

KIGUCHI: That's right. Newton's discovery of universal gravitation was the greatest of all scientific discoveries up until the seventeenth century. The discovery of the nuclear force has been called the victory of twentieth-century physics. In this century, our understanding of universal forces has advanced, and, at the same time, the vast energies inherent in the infinitely tiny world of atomic nuclei and elementary particles have been discovered through the work of Dr. Yukawa and others. Through new disclosures about of the nature of time and space, the existence of this hitherto unimaginable source of energy has become increasingly clear.

IKEDA: Putting it simply, could we say that these new discoveries emerged from Einstein's theory of relativity and from the quantum theory?

KIGUCHI: Yes. The universal applicability of these two theories has been corroborated by experimentation.

IKEDA: Newton's discovery of universal gravitation, prompted by his observation of a falling apple, strikes us as something we can more or less understand. But, from our perspective, Einstein's theories seem difficult to grasp.

KIGUCHI: Perhaps it is because, while Newton's theories deal with the interactions among visible bodies, Einstein clarified the vast potential energy that exists within visible matter but cannot be seen. According to his calculations, the energy inherent in matter is equal to its mass times the square of the speed of light. It is a universal principle, which explains that even a small mass, when converted to energy, yields a tremendous amount of power. In the scientific world, until that time, it was thought that the total amount of matter and the total amount of energy in the universe would remain eternally constant. To apply a numerical value to Einstein's equation $E=mc^2$, one gram of mass equals approximately twenty trillion calories. This means that one gram of mass can warm one million cubic meters of water by 20°C.

SHIMURA: Are nuclear weapons derived from this principle?

KIGUCHI: That's correct. When one atomic nucleus is converted into another by means of a nuclear reaction, because the strength of the nuclear force changes, the atomic weight of the original substance also changes. The weight difference before and after the reaction equals the energy released. The weight of the atomic nucleus per nucleon (proton or neutron) of iron is the lightest of all the other elements. When a nucleus is converted into another, lighter nucleus, in terms of per-nucleon weight, the product of the reaction is energy. The reaction changing from hydrogen into heavier elements toward iron[3] is called nuclear fusion, on which principle hydrogen bombs are produced. On the other hand, the reaction changing from uranium into lighter elements toward iron is called nuclear fission, which is the principle used for the production of atomic bombs. Nuclear reactions, by which matter is transformed into great amounts of energy, is known to be the source of energy that generates the light of stars. On Earth, nuclear reactions hardly ever occur in nature, and have mostly occurred through human design.

IKEDA: I've heard that the bomb dropped on Hiroshima equalled the amount of energy contained in a single gram of matter. Humanity now holds in its hands truly formidable amounts of cosmic energy, capable of utterly destroying this irreplaceable 'oasis of life' that is our Earth. People can no longer be indifferent toward peace, but must strive to act with wisdom to bring about its realization.

SHIMURA: We have indeed entered such an age. I hear there are more than fifty thousand nuclear warheads stockpiled worldwide. This is said to equal 1.6 million times the destructive power of the bomb dropped on Hiroshima. Moreover, such stockpiling of arms continues to escalate.

IKEDA: It brings to mind a passage by Nichiren Daishōnin which states, 'The wise may be called human, but the thoughtless are no more than animals.'

SHIMURA: Humanity has reached the point where it can no longer control this monster that has been created.

KIGUCHI: The diabolical nature of nuclear weapons undermines civilization just as cancer cells destroy the life of an individual – by the time one becomes aware of its presence, the illness may have already become terminal.

SHIMURA: Among the destructive effects, we find the prevalence of a despairing attitude, that since everyone eventually has to die anyway, there is no point in trying to do anything.

IKEDA: Generally speaking, it is hard for people in a relatively tranquil state to really feel the necessity for peace, or to appreciate how great a blessing peace is. In a similar way, it is difficult to sense the horror of a nuclear holocaust.

SHIMURA: That's true. For example, we hear the irresponsible argument that, since the cities of Hiroshima and Nagasaki were rebuilt in less than ten years, even if a nuclear war were to happen, recovery would still be possible.

KIGUCHI: A fundamental solution, I believe, does not lie in this or that methodology or in detailed discussions of the issues, but rather in the willingness of each nation to confer with others for the sake of peace, and in a joining of the voices of the common people desiring peace.

Sitting on a Powder Keg

SHIMURA: By the way, the American film, *The Day After*, depicting the tragedy of nuclear war, seems to have been well received the world over.

IKEDA: I saw a video of it. It included scenes that can only be called horrifying.

KIGUCHI: They showed vividly what would happen to ordinary citizens in the event of a sudden nuclear attack.

SHIMURA: Even the White House was unable to overlook the potential influence of this film. It seems they even felt it necessary to announce at a press conference that, so far, it had had no negative political impact on the president.

SHIMURA: That movie conveys a number of elements which underline the tragedy of nuclear warfare. For example, up until the moment when the nuclear bomb actually explodes over the town, its inhabitants continue to believe that such a thing could not possibly happen in so peaceful a spot. Similarly, once the order for nuclear attack had been confirmed, everything was completely out of human hands, and was carried out in a very businesslike and mechanical fashion.

IKEDA: Altogether, it was a shocking film. But I believe the real horror of nuclear war is not the things you have mentioned. The indescribable torment and grief of the victims can never be fully captured on film.

SHIMURA: Dr. Carl Sagan, when he spoke with you in 1983, remarked that the tragedy of nuclear warfare far surpasses anything that can be shown on film.

KIGUCHI: The Americans lead their lives surrounded by nuclear missiles. In fact, we have reached a point where all people of the world live surrounded by nuclear missiles. The entire human race is sitting on a powder keg, so to speak.

SHIMURA: That is all too true. Actually, among the leading figures in the United States who viewed this film, some clearly referred to what Mr. Ikeda has just pointed out. For example, Dr. Henry Kissinger apparently didn't care for it, remarking that the film presents a very simple-minded view of the nuclear problem.
　Mr. Ikeda, you conferred with Dr. Kissinger on several occasions, didn't you?

IKEDA: Yes, three times, here and in the United States. We talked about peace, culture and other subjects. I recall his telling me earnestly that peace means more than the absence of war, and of his belief that philosophy, or perhaps religion, is absolutely vital to the peace effort.

SHIMURA: In recent times, some people have actually argued the possibility of waging 'limited nuclear war' with neutron bombs, thus attempting to rationalize the use of nuclear weapons.

IKEDA: A generation ago, we heard the ridiculous argument that the fearful nature of nuclear weapons in and of itself would act as a mutual brake on war. Now, this unrealistic logic has completely collapsed, due to the emergence of 'usable nuclear weapons', that supposedly make it possible to attack one's enemies alone, without inflicting direct harm to other nations.

SHIMURA: World-wide military expenditures continue to escalate, and we are beginning to see the development of orbiting offensive weapons, laser weaponry and the like. Mutual mistrust among nations has brought about unlimited arms expansion.

IKEDA: For that very reason, our times, more than ever before, demand that we clarify the diabolical nature of nuclear weapons on a world-wide scale.

SHIMURA: The anti-nuclear exhibit entitled 'Nuclear Arms: Threat to Our World', sponsored jointly by the United Nations Department of Public Information (UNDPI), the Sōka Gakkai, and the cities of Hiroshima and Nagasaki, seems to have elicited a tremendous response.

IKEDA: Yes, indeed. Thanks to the young people's efforts and passionate desire for peace, this exhibit was displayed in New York, Geneva, Vienna, Paris and in other parts of the world.

KIGUCHI: The exhibit must have been especially shocking to Europeans and Americans, who have had no experience of atomic destruction.

SHIMURA: I would imagine so. Sōka Gakkai members carried out a ten-million-signature petition campaign for the total abolition of nuclear weapons. They have also displayed an anti-war, anti-nuclear exhibit seen by about 1.3 million people in forty-four locations throughout Japan, and have compiled no less than seventy-eight volumes of anti-war publications, relating people's personal experiences of the tragedy and brutality of war.

KIGUCHI: I hear some of these have been published in several foreign languages.

IKEDA: Yes, in three volumes so far.

SHIMURA: Another Sōka Gakkai-sponsored touring exhibition titled 'Actions for Peace' was exhibited in January 1984, in Tochigi Prefecture, Japan. Displayed so far in twenty locations, it has attracted a total of 1.1 million viewers. Everywhere it has been exhibited, prefectural governors as well as scholars and cultural leaders have been deeply impressed.

The Sōka Gakkai has also established five peace exhibition halls – in Yokohama, Osaka, and other localities – where materials related to war and peace have been assembled and put on display. So far, about six hundred thousand visitors have seen these exhibits, including Sōka Gakkai members.

KIGUCHI: When we consider the Sōka Gakkai's other activities, such as aid to refugees and the sponsoring of public lectures on a wide variety of subjects related to peace, we can see the broad scope of its endeavours.

SHIMURA: Actions and exchanges for the sake of peace conducted at a grass-roots level, which Mr. Ikeda has proposed, are now being carried out both at home and abroad thanks to the efforts of young people.

IKEDA: It is gratifying to hear you say so. I feel it is our first obligation to bequeath a peaceful world to the next generation.

SHIMURA: I agree with you wholeheartedly. It is easy to talk about peace, but all too difficult to achieve it. Which reminds me, some time ago I heard a lecture on the nuclear issue by Yoshiki Hidaka, the Washington DC bureau chief of NHK (Japan Broadcasting Corporation).

KIGUCHI: The words of a first-class journalist sometimes

carry more immediacy and persuasive power than those of a scholar.

Shimura: Mr. Hidaka has spent more than ten years in Washington, the political nerve centre of the world. Speaking from his wealth of experience, he said that actions for peace and grass-roots diplomacy, such as those conducted by Sōka Gakkai International President Daisaku Ikeda, who has repeatedly conferred with many national leaders, provide an important key to breaking our present deadlock.

Ikeda: I'm embarrassed by your excessive praise. More than ten years ago, I believe, I met him in Shizuoka with his wife and two charming daughters. It was at the height of the cherry-blossom season, and the trees were raining down petals on our shoulders. Even now I remember it vividly.

Shimura: Leaders can no longer confine themselves solely to the dimension of ideology, ethnic concerns, or politics. Freely transcending the concept of national boundaries, and without calculation for self-interest, they must strive single-mindedly for the peace and stability of the human race as a whole. People are eagerly awaiting the emergence of such leaders, don't you think?

Kiguchi: I agree. We have indeed entered an age when such leadership is vital.

Shimura: Mr. Hidaka also said that such leaders are certain to be noticed and appreciated by educated people in countries where democracy is well established.

Ikeda: I hope that many such advocates of peace will emerge in each nation. Personally, based on my conviction as a Buddhist, I am determined to exert myself to the utmost, travelling to the farthest corners of the Earth, in order to try and help resolve the fundamental causes of human suffering – the problem of birth and death – as well as fundamental illusions such as greed, anger, ignorance, arrogance and doubt.

SHIMURA: A philosophy of peace at the political level is important, but will prove inadequate without a corresponding philosophy of peace at the human level. The reason is that the power to resolve one's individual suffering, and the experience of having done so, will, in itself, help generate a rushing current leading toward the solution of the sufferings of society as a whole.

KIGUCHI: You are absolutely right. No matter how a person may raise the cry for the peace of humanity, even if he is respected as an outstanding scientist or scholar, if he himself is trapped by his karma, tormented, perhaps, by family discord, what power can his words ultimately carry?

SHIMURA: To come down to specifics, if we are to protect human life, which is more precious than the Earth itself, the abolition of nuclear weapons is one issue we cannot avoid.

IKEDA: I believe that is true. We have seen many movements to abolish nuclear arms, led by Nobel Prize winners and other outstanding individuals from various fields. Each of these efforts has had its own value and impact. However, with the passage of time those who at first had an interest in these movements become distracted by the demands of daily life. As a result, such movements tend to fall into oblivion. In essence, I think too many people tend to feel that someone else will address this problem, and thus fail to recognize it as their own personal responsibility.

Jōsei Toda's Anti-Nuclear Weapons Declaration

SHIMURA: Jōsei Toda, second president of the Sōka Gakkai, made a famous speech in which he pointed to the diabolical nature of nuclear weapons.

KIGUCHI: With his unceasing desire that nuclear weapons will never be used again no matter what, he went so far as to declare that those who employ such weapons deserve the death penalty.

SHIMURA: He made that speech at Mitsuzawa Stadium in Yokohama, on 8 September 1957, before an audience of fifty thousand young people.

IKEDA: That's correct. President Toda's declaration calling for the total abolition of nuclear weapons, which was his will, never leaves my mind, even for a moment. That is because he enjoined us, as his followers, to carry it out. No matter who may criticize or slander me for my efforts, I have no fear. From the standpoint of my sense of responsibility to humanity, I believe that my path is a correct one.

SHIMURA: But why did President Toda, a Buddhist, make the extreme statement that those who employ nuclear weapons should receive the death penalty?

IKEDA: It is indeed a strong statement. However, with the passage of time, I have come to feel that it carries a profound meaning.

SHIMURA: Some have criticized his remark half-jokingly, saying that if nuclear bombs were to be dropped, there would be no one capable of carrying out the death sentence.

KIGUCHI: I've heard that one of the crew members of the bomber that dropped the atomic bomb over Hiroshima later lost his sanity.

SHIMURA: Our monthly journal *Ushio* once published his memoirs, by exclusive arrangement. Actually, you are speaking of Claude Eatherly, who flew advance weather reconnaissance for the bomb plane, the B29 *Enola Gay*, and witnessed the dropping of the atomic bomb. It is said that he became mentally deranged and died, completely destroyed by what he had seen.[4] Another member of the flight crew entered the religious life.

IKEDA: Eatherly left behind his famous correspondence with a psychiatrist, entitled *Burning Conscience*.[5] In any event, we can say that having dropped an atomic bomb, in and of itself, constitutes a hell far worse than the death penalty.

KIGUCHI: It would certainly seem so.

IKEDA: The intent to employ these fearful weapons is nothing other than a function of life's inherent diabolical nature. I say this because it stems from a mind that would incinerate hundreds of thousands, or even millions, of human beings in a single moment. Perhaps Mr. Toda ventured to use the words 'death penalty' to mean that a mind steeped in what Buddhism terms illusion or ignorance must 'die' – that is, must be transformed into the mind of enlightenment.

SHIMURA: That was the same year as the famous Russell-Einstein statement against nuclear weapons and the first Pugwash Conference.

KIGUCHI: That conference was attended by scientists from all over the world, and manifested their desire to totally ban all nuclear weapons and to banish war from the face of the Earth.

SHIMURA: They stood up and declared that scientists throughout the world must pool their abilities to accomplish peace.

KIGUCHI: However, contrary to their desire, nuclear weapons have continued to proliferate. Nor does the presence of these weapons seem to have served as a decisive brake in preventing the wars that have followed.

SHIMURA: At that time, the United States and the European nations were, on the whole, not yet fully aware of what the nuclear menace implied.

IKEDA: Therein, I believe, lies the significance of the Pugwash Conference. It is truly regrettable that it failed to evolve into a world-wide movement encompassing the great mass of the common people.

SHIMURA: Just two months after that conference, Mr. Toda earnestly appealed to the young people who should shoulder responsibility for the future, saying, 'I hope that you, who hear my declaration today, will carry it out and make its meaning known throughout the world.'

KIGUCHI: Could one say that Mr. Toda, at that time, foresaw the arrival of an age when peace movements could not afford to be limited to certain groups, but would have to encompass the common people as a whole?

IKEDA: One may certainly view his declaration in that light. A 'fortress of peace' must be built in the lives and minds of individual human beings. Within the individual human being, we find a tendency to create evil karma. On the other hand, we also find the mind that desires, like the bodhisattva, to contribute to the welfare of others. I believe it is appropriate to view Mr. Toda's declaration as a statement stressing the urgent need to triumph over evil karma or the diabolical nature lurking within human life. The more the ranks of common people increase, holding aloft the banner of humanity and raising the cry for peace, the closer the realization of peace will become, and the further the possibility of war will recede into the distance. No doubt Mr. Toda acutely perceived that, without creating wave upon wave of individuals committed to this cause, a total victory for humanity cannot be won.

SHIMURA: That makes sense. It was only twelve years after the Second World War, and Japan was still impoverished in both the material and the spiritual sense. Appeals for peace made at so lofty a level must have seemed remote indeed. In those days some people might even have regarded it as a form of self-advertisement. Mr. Toda's foresight was penetrating, I believe. I recently read a book entitled *Nihon no ikikata to heiwa mondai*[6] (*The Japanese Way of Living and the Issue of Peace*). It explores the question of how a peace movement at the grass-roots level might be generated. It also discusses the universal application of the spirit of Article Nine of the Constitution of Japan.[7]

IKEDA: I haven't read it yet. I truly regret the fact that lately I simply don't have much time for reading, probably because of my hectic schedule. You two are still young, however, so I hope you will read widely. Shortly before Mr. Toda's death – I believe it was just two weeks before he passed away – he startled me by suddenly asking, 'What have you read today?' Moreover, he was always saying, 'You young people must read continually and teach me what you learn.' I cannot forget his words. And now I've reached the age that he was then. Be that as it may, I believe that the desire to renounce war has already become deeply rooted in the hearts of people the world over. And, having taken root in the hearts of the people, it must inevitably sprout and grow to form a trunk, leaves and branches, and eventually bear the fruit of the abolition of war and the establishment of eternal peace. Many people throughout the world may not know that, in addition to Japan's being the only country in the world to have experienced an atomic attack, the spirit of perpetuating peace is extolled in Article Nine of her Constitution. However, I feel our fellow citizens must fully awaken to the increasing importance of Japan's mission for peace.

KIGUCHI: I feel we are standing at a major crossroads that will determine whether or not Japan can walk the path of a peaceful nation, trusted by the rest of the world.

IKEDA: Speaking from an even more profound perspective, as a Buddhist, I feel there must be a specific reason why Nichiren Daishōnin was born in Japan and proclaimed the dignity of life.

SHIMURA: Incidentally, the book I mentioned a moment ago quotes the Japanese scientist Toshiyuki Toyoda, in reference to the issue of peace and nuclear weapons, as saying, 'Ultimately, this is a problem of the human being.' That greatly impressed me.

IKEDA: Then he is in accord with what we have been continually stressing for the past few decades.

Good and Evil Inherent in Life

SHIMURA: A Dutch philosopher stated, 'Peace is not an absence of war; it is a virtue, a state of mind, a disposition for benevolence, confidence, justice.'[8]

IKEDA: Those words of Spinoza are quite famous. I fully agree with their sentiment. No matter how passionately a person may raise the cry for peace or even strive for its achievement, without some sound philosophy or way of thought to elevate his own humanity, any advance toward peace that one may accomplish will not endure.

KIGUCHI: Benevolence and justice stand in direct contradiction to human desires and cravings – which Buddhism terms *bonnō*. As long as one ignores the difficulty of liberating oneself from these drives, genuine self-improvement and the reform of society will prove impossible.

SHIMURA: That is a crucial point. What is the actual meaning of *bonnō*, earthly desires or illusions?

IKEDA: To explain simply, Buddhism teaches that human beings are subject to three categories of illusion: illusions of thought and desire, illusions innumerable as particles of dust and sand, and illusions about the true nature of existence. The illusions of thought and desire, as the name indicates, actually comprise two categories. Illusions of thought are false views, or distorted perceptions of the truth of things. They are regarded as mental and learned. In contrast, the illusions of desire refer to emotional, instinctual illusions that one is born with, such as greed, anger, stupidity, arrogance, and so forth.

SHIMURA: Then it would be correct to understand *bonnō* – translated as illusions or earthly desires – as a generic term for all the spiritual functions that trouble us in both body and mind, is that correct?

IKEDA: I believe we may define *bonnō* in that way. However, because they are functions of life itself, one cannot eradicate them, even if one tries.

SHIMURA: What is meant by 'illusions innumerable as particles of dust and sand'?

IKEDA: Simply stated, the illusions of thought and desire which I just mentioned are illusions arising within oneself. In contrast, we may think of the 'illusions innumerable as particles of dust and sand' as those that arise when one attempts to save others. In order to save people of varying characters and situations, one must understand all sorts of things in the social dimension. The illusions impeding that understanding belong to this category. Or, from another standpoint, no matter how many relief measures or welfare systems one may initiate in society, unless they are grounded in the spirit of compassion, those who are not helped thereby will remain 'innumerable as particles of dust and sand.'

KIGUCHI: What about illusions concerning the true nature of existence?

IKEDA: These are illusions springing from what Buddhism calls *mumyō*, or darkness. 'Darkness' may be thought of as the function which seeks to destroy all that sustains and enriches life. We can say that 'darkness' indicates a state in which all the human being's inherent life-affirming powers, all his potential for boundless wisdom and compassionate actions, are completely blocked. Concerning this state, Nichiren Daishōnin states in the *Sōkanmon shō* (*On the Teachings Affirmed by All Buddhas throughout Time*), ' "Darkness" means inability to penetrate. It indicates not having clearly awakened to the true nature of one's life.'

Moreover, the illusions about the true nature of existence are also called fundamental illusions, meaning that the illusions of thought and desire, as well as the illusions innumerable as particles of dust and sand, ultimately derive from this category. In showing how firmly rooted such illusions are, the Daishōnin states in the *Genkō gosho* (*On the Declining Kalpa*), 'The greed, anger and stupidity in people's hearts in this degenerate latter age are such that even the greatest wise man or sage could not subdue them.' Nichiren Daishōnin's Buddhism teaches a way, which is concrete, whereby this incorrigible 'darkness' innate in life can be transformed into enlightenment.

SHIMURA: May we take this to mean that anyone can transform darkness into enlightenment through the power of the Mystic Law?

IKEDA: Yes. As the *Ongi kuden* states, '*Myō* indicates enlightenment, and *hō*, darkness. Darkness and enlightenment are [two aspects of] one entity; this is called the Mystic Law [*myōhō*].' This topic may be somewhat difficult, but what is called Mahayana Buddhism teaches that it is not necessary to eradicate the illusions and desires human beings possess. To do so would in fact amount to the denial of life itself.

KIGUCHI: Does this indicate, then, that human beings, tortured as they are by problems, suffering and grief, also possess the most respectworthy, essential nature of the Law?

IKEDA: That's correct. Mahayana Buddhism therefore firmly rejects the unrealistic, life-denying practices expounded by Hinayana Buddhism, referring to them as disciplines which call for 'reducing one's body to ashes and annihilating one's consciousness'.

SHIMURA: Then Mahayana Buddhism does not deny earthly desires?

IKEDA: That's correct. Nichiren Daishōnin states in the previously cited *Jibyō shō*, 'The heart of the Hokke [Lotus] sect is the principle of *ichinen sanzen*, which reveals that both good and evil are inherent even in those at the highest stage, that of *myōgaku* or enlightenment. The fundamental nature of enlightenment manifests itself as Bonten and Taishaku, whereas the fundamental darkness manifests itself as the Devil of the Sixth Heaven.' Put concisely, the stage of *myōgaku* or enlightenment indicates the enlightened life-state of the Buddha. This passage teaches that even this state of life is endowed with the nature of both good and evil. All existences in the universe possess the nature of both illusion and enlightenment. In the microcosm of the human individual, these express themselves as the workings of good and evil. When one bases himself upon the Mystic Law, his life will manifest the functions of Bonten and Taishaku, who represent all the workings that act to protect human existence.

SHIMURA: What is indicated by the Devil of the Sixth Heaven?

IKEDA: We can think of this as the source of all destructive functions of human existence. This devil is said to dwell in the uppermost heaven of the world of desire. Because he delights in sapping the life force of others, he is also called the 'robber of life'. In terms of the individual, this fundamental, diabolical nature robs one of the power to support and enhance one's life. In terms of society, it works to deprive a great many people of their lives. For that reason, I believe we can regard nuclear weapons as an expression of what Buddhism terms the Devil of the Sixth Heaven. However, we are endowed with the power to suppress and control this diabolical nature. Because 'darkness and enlightenment are [two aspects of] one entity,' this nature and earthly desires can be elevated to a lofty dimension, which truly benefits human beings. This concept is explained by the principles that earthly desires are in themselves enlightenment (*bonnō soku bodai*), and darkness is in itself enlightened nature (*mumyō soku hosshō*). In this connection, Nichiren Daishōnin states that 'those who chant *Nam-myōhō-renge-kyō* . . . burn the firewood of earthly desires and behold the fire of enlightened wisdom before their eyes.' When his life is illuminated by the Mystic Law, the individual 'burns' his earthly desires in his efforts to contribute to the welfare of others and to society, while on the level of society as a whole, leaders 'burn' their earthly desires in their efforts to ensure the happiness of the people and lead them toward peace.

SHIMURA: Then earthly desires can be transformed into tremendous energy to save others from unhappiness.

KIGUCHI: This is Buddhism's superlative view of life, which our ordinary understanding cannot possibly fathom. It can also be called a great principle of life-reformation.

IKEDA: Indeed, though it may seem like the long route, this steady path of reformation within individual human life constitutes the first sure step toward peace.

KIGUCHI: I completely agree. In that sense, I feel drawn to a way of living based on the Middle Way of Buddhism, which embodies both humanism and reverence for life.

Blocking the Path to Destruction

SHIMURA: Mr. Ikeda, you recently had a dialogue with Professor John D. Montgomery of Harvard University, an authority on the issues of peace and development. I hear that he evaluated the Sōka Gakkai's peace movement highly, saying in contrast that peace movements which overlook the essential nature of the human being ultimately prove unproductive, while those which affirm it are constructive and viable.

IKEDA: Yes, that was a most enjoyable discussion. Professor Montgomery, of the political science department at Harvard, is a very humble and affable individual. He has the sharp eyes of a scholar. Dr. Glenn D. Page of the University of Hawaii, a scholar involved in peace issues and a friend of mine for many years, also took part in that discussion. I believe he was just about to depart for a meeting of the World Future Studies Federation in Stockholm, but made time to attend. In any event, we have seen revolutions in the arenas of both politics and economics. Industry and science have had their revolutions as well. However, despite their undeniable gains, these revolutions have invariably brought in their stead new problems of their own. I, therefore, believe that the last revolution left to humanity is a reformation of the spirit – what we call the human revolution. That is why we are always asserting that the way to block the path to destruction lies in our movement, which is based upon Buddhism and regards the human being as the basis of all endeavours. Nichiren Daishonin teaches us the precise way to carry out this reformation when he states, 'Base your heart on the ninth consciousness and your practice on the six consciousnesses.'

SHIMURA: What are the six consciousnesses?

IKEDA: Put simply, we can describe them as six divisions of the functions of consciousness, such as the perceptions of the sense and the workings of the mind. The first four are sight, hearing, smell and taste, the fifth is touch, and the sixth is mind. We can view these six consciousnesses as the functions that enable us to carry out our day-to-day human affairs, for example, in the family, in society, and so on. The Daishonin teaches that faith must manifest itself in this realm of daily life. Ours is not a Buddhist practice wherein you seclude yourself on a mountain top or in some remote valley. Nor do I believe that one can carry out faith in total isolation from others.

KIGUCHI: However, I understand that earlier forms of Buddhism did set forth the enlightened state of life as a goal remote from one's immediate reality.

IKEDA: In the provisional Buddhist teachings, that was definitely the case. To that end, they emphasized austere practices spanning many lifetimes in order to reach the goal of 'enlightenment'. If one is limited to this view, however, practice will fail to transcend the realm of personal gratification. Moreover, such doctrines provide no basis for the reformation of society.

SHIMURA: That's true. Religion should not be an abstract pursuit.

IKEDA: Nichiren Daishonin's Buddhism embodies the Mystic Law or *Nam-myōhō-renge-kyō* in the form of the object of worship. He taught that, by simply chanting *Nam-myōhō-renge-kyō* to this object of worship, one activates the powers of the

Buddha and the Law inherent in his own life, and based on the principle of the simultaneity of cause (the nine worlds, or the life of the common mortal) and effect (Buddhahood), one immediately gains access in reality to the most profound and sacred dimension of life, called the ninth consciousness.

In short, to 'base one's heart on the ninth consciousness' means to continuously orient *oneself* to a life-state awakened to the ninth consciousness by believing in and chanting *Nam-myōhō-renge-kyō*. *Nam-myōhō-renge-kyō* is itself 'the ninth consciousness, the unchanging reality which reigns over all life's functions, and is inherent in our own lives.

Moreover, to 'base one's practice on the six consciousnesses' means to act positively in the realities of daily life and society by basing this action on the ninth consciousness. It means to act so as to contribute to others' happiness and to establish a peaceful world. Moreover, it means to provide the guidelines, so that many others can also lead their lives harmoniously. In short, it indicates the building of a society wherein people can live securely and lead rewarding lives. I believe that it points to our activities for the establishment of a Buddha land, or truly peaceful and happy societies.

SHIMURA: It indicates that the actual flowering of the individual should take place within the context of the times and society in which he lives; it is not mere idealism.

IKEDA: In this connection, the Mystic Law will restore boundless creative energy to people in our present, deadlocked society. That is why it is called the Law which 'revives' people, because it is a source of limitless vitality. A movement of awakened individuals, based on this Law, will generate a reliable force for peace and create momentum in that direction.

KIGUCHI: A true peace movement is based on continual self-improvement and a life of fulfilment.

SHIMURA: With this firmly in mind, we devote ourselves wholeheartedly to our work, raise our children, support our families, and study diligently. Efforts should be carried out in the midst of daily life to spread an unwavering 'will toward peace' among the people. In fact, without this emphasis on daily life, our peace movement cannot last.

The other day I heard of the following incident from a friend of mine who has resided for many years in Europe and was making a brief trip to Japan. It seems that he had met Mr. Rikhi Jaipal, a UN assistant secretary-general in Geneva, who had apparently commented that, up until now, many of the anti-nuclear and peace movements have also tended to be anti-authority and anti-government. In contrast, the Sōka Gakkai's peace movement is rooted in belief in the dignity of human life, and has stimulated a broad awakening on the grass-roots level. In this sense, he said, it differs markedly from other such movements and drew his heartfelt approval. Moreover, I understand he expressed his desire to support our efforts in the future.

KIGUCHI: Many sensitive people are beginning to notice our movement, which focuses on the human being, and are discovering that it is a sound basis for hope.

NOTES: CHAPTER XI

1 Oda Nobunaga: One of the major feudal warlords during the period of civil wars which took place between the mid-15th century and the mid-16th century, who was greatly interested in Western learning.
2 Tokugawa Ieyasu: He succeeded in unifying Japan and established the Tokugawa shogunate in 1603.
3 In this case, although the reaction changes into 'heavier' elements in the periodic table toward iron, the weight of the nucleus per nucleon becomes lighter.
4 *See* Robert Jay Lifton, *Death in Life, Survivors of Hiroshima* (New York: Random House, 1969).
5 *See* Claude Eatherly and Gunther Anders, *Burning Conscience* (New York: Monthly Review Press, 1962).
6 Published by Iwanami Shoten, Tokyo.
7 The no-war, no-arms provision of the Constitution of Japan, which reads: Article 9. Aspiring sincerely to an international peace based on justice and order, the Japanese people forever renounce war as a sovereign right of

the nation and the threat or use of force as a means of settling international disputes.

In order to accomplish the aim of the preceding paragraph, land, sea, and air forces, as well as other war potential, will never be maintained. The right of belligerency of the state will not be recognized.

8 Baruch Spinoza, *Theological-Political Treatise*, 1670.

XII

Does the Universe Undergo Birth and Death?

The universe from which all matter and energy are derived: the closer we are to understanding its true aspect, the more we can recognize the penetrating insight of the Buddhist teachings.

'Heavenly Beings' in Spacesuits

SHIMURA: Recently an astronaut successfully completed the first attempt at extra-vehicular activity without an 'umbilical' cord attaching him to his spacecraft.

KIGUCHI: Yes, I watched it with great excitement on the news. Since there is no gravity in space, even if an astronaut steps outside his ship, he will not fall. However, it is very difficult for him to manoeuvre. That's why it's dangerous. In order to move about, he carries a tank of nitrogen gas fastened to his back, and manoeuvres himself by jet propulsion. It's as though the astronaut becomes a miniature rocket.

SHIMURA: That must be why he looks as though he's sitting in a chair.

IKEDA: On temple walls, one often sees murals that depict heavenly goddesses clad in robes, soaring freely through the skies. Now, toward the end of the twentieth century, astronauts walk through space like 'heavenly beings' clad in white spacesuits. Though, I must admit, there's a slight difference in style (laughter).

SHIMURA: In any event, space is becoming increasingly accessible to us.

IKEDA: That's quite true. In an age like ours, human beings can no longer afford to cling to a limited perspective or narrow conceptions of the world. The times must move in the direction of world peace, toward a world without war. I hope that national leaders in particular will formulate their policies with this end in view.

SHIMURA: Not long ago, when I was in Europe, I had the opportunity to visit Mr. Robert Maxwell, the president of a famous British publishing firm. He told me that he had been involved in the planning of the Pugwash Conference, the first occasion on which world scientists joined together and spoke out in the cause of peace. He also told me that he had, until recently, thought science and religion were moving farther apart, as time passes. He was astonished to learn that we were discussing religion in the light of the latest findings of modern science.

KIGUCHI: This is probably the first time that he has heard mention of modern scientific issues in connection with Buddhism.

IKEDA: I believe that the role to be played by religion in the scientific age will become increasingly significant.

KIGUCHI: I recall the passage from the Russell-Einstein statement against atomic weapons which reads in part: 'We appeal, as human beings, to human beings: remember your humanity, and forget the rest. If you can do so, the way leads to a new Paradise; if you cannot, there lies before you the risk of universal death.' As a scientist, this impressed me deeply.

SHIMURA: It brings to mind the fact that Einstein, in his later years, expressed his expectations of the wisdom of the East to play a great part.

IKEDA: A great scientist will, after all, strive to develop great humanism. I feel that, in order to do so, he cannot help but turn to religion, and in particular, to Buddhism, which is the pinnacle of Eastern philosophy.

SHIMURA: By the way, Mr. Ikeda, I hear that you received an extraordinary gift from NASA.

IKEDA: When I met Dr. Carr, the *Skylab 4* commander in 1983, as an expression of friendship, he gave me, on loan, specimens of six types of rock found on the Moon and six types of meteorite fragments collected from countries throughout the world. Since I want as many young people as possible to see them, they are now on display in a display room relegated for Sōka University's use within the Tokyo Fuji Art Museum in Hachiōji. Dr. Carr also presented us with a special NASA film.

Not long ago (30 January 1984), I had it shown to the students of Sōka Junior and Senior High Schools and Sōka Primary School in Tokyo. Perhaps it is simply the natural curiosity of youth, but these children appear positively hungry for knowledge. They seemed tremendously excited by their first visual experience of space.

SHIMURA: Mr. Ikeda, I've noticed that you invite many important foreign guests, not only to Sōka University, but also to the Sōka schools as well.

IKEDA: That's true. Offhand, I can think of the late Rector Rem Khokhlov of Moscow State University, Rectors Juan de Dios Guevara and Gaston Pons Muzzo of the University of San Marcos in Peru, Mr. René Huyghe from France, and a number of scholars and friends from China. No matter how strongly one may appeal to young people and stress the need for cultural exchange in order to promote peace, without concrete action, it remains an empty cause. Therefore, in this respect, I wish to treat children and young people in exactly the same manner as adults, and create as many opportunities as possible for them to participate in cultural exchanges with people from different countries.

KIGUCHI: It seems that visitors to the schools have been quite impressed by the earnestness and diligence of the pupils. It never fails to impress me that, despite your busy schedule, you always make time to talk with students and other young people.

SHIMURA: Yukichi Fukuzawa (1835-1901), a great educator of the Meiji period and one of the first Japanese to travel widely abroad, sought out the company of his students whenever possible and often told them that being together with their teacher is the same thing as sharing his knowledge.

IKEDA: It is truly regrettable that educators such as he have become rare. Whether eating, or taking a public bath, or gazing up at the stars in the night sky, leaders should take as many opportunities as possible to talk with young people about life and about peace. I feel this effort is very important. I recall once having had a lively discussion with some students in a community bath house. They told me much later that even after they had grown up and taken their places in society, they cherished our conversation as a precious memory, although I didn't mean to make it a special occasion. I keenly feel the need for direct contact with the younger generation.

Observation of Mars

SHIMURA: By the way, Dr. Kiguchi, I hear that Mars is approaching the Earth in its orbit around the Sun.

KIGUCHI: That's correct. This time the Earth and Mars will reach their closest points in their respective orbits on 19 May 1984, at about 8.00 in the evening.

IKEDA: Will Mars appear much larger than usual?

KIGUCHI: To the naked eye, it probably won't appear any different than usual. However, magnified about a hundred times by a telescope, it will seem like the full moon when viewed with the naked eye. However, because of Mars's elliptical orbit, the

Sun, the Earth and Mars will come into alignment on 11 May, at about 5.00 pm

SHIMURA: Mars's approach this time is a minor opposition, which occurs about every two years; it is not the most favourable opposition, is it?

KIGUCHI: Right. The most favourable opposition occurs only once every fifteen or seventeen years, the next one being in September 1988.

IKEDA: How large will Mars appear at that time?

KIGUCHI: About as big as a coin seen with the naked eye from two hundred to three hundred metres. The planet will appear reddish-orange.

IKEDA: In Japanese Mars is called *kasei*, the fire star. We find the 'fire star' mentioned in Nichiren Daishōnin's writings. The word *kasei* appears in the *Risshō ankoku ron*, the writing famous for the Daishōnin's prediction of the Mongol invasion.

SHIMURA: Actually it appears in a passage cited in that treatise from the *Ninnō Sutra*. People at that time generally referred to Mars as *keiwaku* (Ch *ying hue*: the ghost star). That is, for example, how the Chinese historian, Ssu-ma Ch'ien, refers to it in his *Records of the Historian*.

IKEDA: We also find this designation 'ghost star' in the *Nihon shoki* (*Chronicles of Japan*), said to be one of the earliest written records in Japan. According to materials concerning the history of astronomy, the term 'fire star' appears in 1824, specifically in the *Yōzō kaiku ron* (*Comprehensive Approach to Universal Nature*), published during the Edo period (1600-1867).

SHIMURA: It seems that the first person to use the characters for 'fire star' to indicate the planet Mars was Kumārajīva, well-known for his translation of the *Lotus Sutra*. He lived about sixteen hundred years ago. In China, even minor changes in the appearance of this planet were thought to be precursors of famine or armed uprisings.

KIGUCHI: In Europe, too, people called this planet Mars, the name of the Roman god of war. Astronomers such as Kepler and Copernicus, noting Mars's elliptical orbit, apparently feared that it might collide with the Earth.

Some astronomers, including the famous Fred Hoyle of Great Britain, have conducted research indicating that changes in the behaviour of heavenly bodies do in fact seem to exert an influence on the Earth and human beings. In any event, the Daishōnin apparently knew a great deal about heavenly bodies.

IKEDA: Buddhism is a teaching which clarifies the fundamental sources of happiness and unhappiness. We can say that it is only natural, in this sense, that it should include profound indications about the relationship of man and the cosmos.

Ryūnosuke Akutagawa's View of Martians

SHIMURA: Mars is said to be about half the size of the Earth. How long did it take to form?

KIGUCHI: The core of Mars formed over approximately two years. It grew relatively rapidly, over a period of about one hundred thousand years, while being bombarded by meteorites. The core of the Earth formed in about one year, but since Mars is farther away from the Sun, it took longer to solidify. Moreover, because it's so far from the Sun, it's a cold planet.

IKEDA: The terrain of Mars is said to resemble that of the Earth.

KIGUCHI: That's true. A mountain on Mars, named Olympus, resembles Mount Fuji to an amazing degree. Its base is extremely broad, about six hundred kilometres.

IKEDA: For a long time – in fact, until receiving the data from spacecrafts which orbited and placed lander modules in the 1970s – it was thought that highly-evolved living beings might exist on Mars. Their presence was conjectured because Mars appears to have water.

KIGUCHI: That's right. Actually, there is water on Mars, but because of the extremely low temperatures, it remains frozen at a depth of one kilometre beneath the planet's surface.

SHIMURA: Sixty years ago, period, the 1924 opposition of Mars caused a great stir in Japan.

KIGUCHI: At that time many people seem to have believed there was life on Mars.

SHIMURA: Even intellectuals, such as the author Ryūnosuke Akutagawa, apparently devoted some thought to Martians.

IKEDA: I recall that Akutagawa conceived of Martians along the lines of features that characterize human beings.

SHIMURA: That view appears in his famous work *Shuju no kotoba* (*Dwarf's Monologue*).

IKEDA: He said something to this effect: When we ask whether or not the denizens of Mars exist, we are really asking whether or not they exist in a form accessible to our five senses. However, life is not necessarily confined to those conditions which our five senses can discern. Martians may possess such a mode of existence that their presence cannot be detected by our five senses. If so, a group of them might even arrive at the Ginza in Tokyo tonight along with the autumn wind which yellows the leaves of the sycamore trees.

With his unrestricted view concerning extraterrestrial life, this passage, along with others, seems to me to suggest that Akutagawa had been deeply influenced by Buddhist thought.

SHIMURA: When speaking of 'Martians', everybody immediately thinks of creatures with long tentacles like octopi. Who first conceived of that idea?

IKEDA: I believe it was the British author, H.G. Wells, in his famous novel, *War of the Worlds*, published some eighty years ago. The image he conceived has become a standard model for later conceptions of extraterrestrials.

KIGUCHI: Astronomers were to an extent responsible for nurturing people's beliefs in the existence of Martians. In 1877, the Italian astronomer G.V. Schiaparelli (1835-1910) was undertaking daily observations of Mars. In the course of these observations, he noticed a number of criss-crossing lines on the planet's surface. He accordingly reported that he had discovered channels on Mars. However, the word 'channels' (Italian *canali*) was mistakenly translated into English as 'canals'. And canals are definitely artificial structures.

IKEDA: So, in other words, there would have to have been intelligent beings who constructed them.

KIGUCHI: That's right. And the more astronomers operated on this false assumption, the more they tended to observe patterns on the surface of Mars that seemed to support it. The strongest exponent of the canal theory was the American astronomer Percival Lowell (1855-1916). He built himself a large observatory in Flagstaff, Arizona, and made faithful sketches of his observations. He found a number of lines criss-crossing the planet's surface in geometric patterns, like the interstices of a net. This must be a network of huge canals, he announced. He hypothesized that Mars had little water, and that the Martians had accordingly built an intricate network of canals to provide water for their survival.

IKEDA: Perhaps the idea that beings possessing such a high level of technology might wish to move to the Earth inspired the science-fiction concept of a 'Martian invasion'.

SHIMURA: It seems as though the image people had of Martians in those days was not entirely unreasonable, given the level of our understanding at that time.

KIGUCHI: No, indeed. For example, Martians were thought to possess large heads, no trunk, and sixteen flexible arms – a very plausible accommodation of the human form to the environment of Mars.

IKEDA: Since the days of ancient Greece, we have seen little change in the ideal conception of the human being, which is to possess a mind capable of contemplation, the capacity to develop civilization to a high degree, and other refined functional abilities.

SHIMURA: That is the ideal image of a living being as conceived by Western civilization.

KIGUCHI: Some biologists believe that wherever it might emerge, intelligent life would inevitably assume a form resembling that of human beings.

The Vessel of the Law

SHIMURA: On the Iranian border lies the Armenian Soviet Socialist Republic. About twenty years ago I was invited there in connection with a Japan-Soviet publication exchange venture. It is surrounded by desert and there is virtually nothing worthy of noting there, but in the past, it appears that Armenia was an important channel for East-West cultural exchanges. Just at that time, an international conference was being held in the capital city, Erivan, on the subject of whether or not intelligent life exists elsewhere in the universe.

KIGUCHI: That was the first conference on such a topic. And it is quite famous for in that reason.

SHIMURA: Some people in the publishing business in Erivan told me about a Soviet biologist's statements at that conference. I've brought some of the notes that I made at the time. Though they are rather long, please allow me to read from them. He said, 'Intelligent life will invariably possess a highly evolved nervous system with a brain at its centre, which must be protected from external danger. It's quite natural that the head should occupy the highest position, where it will have the least weight to carry. Beneath the head, the being will be provided with arms, and it can immediately convert its thoughts into actions. It will require a trunk to house its vital organs. At its lowest point, it will have feet, to provide mobility over the ground. Moreover, its appendages will have to be paired, so that it can balance itself efficiently.'

KIGUCHI: This manner of thinking must be a matter of common sense for biologists. At a more immediate level, we find a number of theories about the reason for the size of the human body and the human head. When we consider the force of the Earth's gravitational pull in conjunction with such factors as chemical bonding power, the human size appears to be optimum indeed.

IKEDA: Buddhism views the human body as the irreplaceable 'vessel of the Law', because human beings alone can practise Buddhism and perceive the ultimate truth within life itself, thereby attaining Buddhahood. I feel we can also interpret 'vessel' in terms of the wondrous human form, apparently well suited to intelligent life anywhere in the cosmos.

SHIMURA: I've heard that if the ice below Mars's surface were to melt, it would form a sea ten metres deep.

KIGUCHI: Yes, but because of the distance between Mars and the Sun, the Sun's rays, upon reaching Mars, do not have enough warmth to melt that ice.

IKEDA: I hear that clouds have been observed around the south pole of Mars.

KIGUCHI: On 20 July 1976, *Viking 1* made a soft landing on Mars. According to the data transmitted from this probe, a phenomenon similar to falling snow was observed.

IKEDA: Mount Olympus on Mars, which you mentioned a moment ago, is a volcano more than two and a half times the height of Everest. Judging from the extent of its lava flows, it appears to have been quite active at one time.

KIGUCHI: Yes, Mars is covered with volcanoes far more extensively than the Earth. More than ten have bases that surpass one hundred kilometres in diametre. Because the gravitational field of Mars is so weak, very massive land forms can take shape without their base collapsing. Thus, mountains reach extraordinary heights. Ten kilometres is said to be about the upper limit for mountains on Earth.

SHIMURA: How much oxygen is there on Mars?

KIGUCHI: Carbon dioxide accounts for more than ninety-five per cent of the atmosphere. Oxygen is estimated to be a mere 0.3 per cent. According to the data retrieved so far from the Martian probe, there are no signs of any living organisms. The planet seems to be covered with reddish brown soil that has a high iron content. The temperature near the equator ranges from sixteen to twenty-four degrees centigrade during the day, though at night it drops to about minus eighty.

SHIMURA: How long does it take to go to Mars?

KIGUCHI: With our present level of technology, about two years.

IKEDA: In December 1983, there was a news report that the American space shuttle *Columbia* photographed a huge Soviet rocket capable of transporting a man to Mars.

KIGUCHI: Yes, apparently, it discovered a rocket base in the Kazakh Soviet Socialist Republic in Central Asia. The rocket you mention is eighty-eight metres long. It is said to have a greater transportation capacity than any other rocket built to date.

Awareness of Death as the Dawning of Religious Consciousness

SHIMURA: One of the most fascinating subjects for anthropological speculation concerns the question of when human beings became aware of the problem of death.

KIGUCHI: It stirs something fundamental in us to learn about our remote ancestors. Your question in effect asks when man first became aware of his humanity.

SHIMURA: That's right. This development is represented by the emergence of the custom of burial.

IKEDA: In the course of human evolution, man's awakening to, and reverence for, the truth inherent in life has found expression in his burial for the dead. I've heard that this custom may date back as far as two hundred thousand years.

KIGUCHI: That would still be in the Stone Age, wouldn't it?

IKEDA: Remains of Neanderthal graves apparently indicate that these ancient people offered flowers for the dead.

SHIMURA: That's correct. About three years ago, an exhibition on 'Neanderthal and Cro-Magnon Man' was held in Tokyo. At that exhibit, the fossilized skeletons of a boy and a girl buried about twenty-five thousand years ago had been reconstructed for display. I was truly impressed by the sight of people stopping to gaze at the exhibit. I felt that the custom of burial, as an act of respect for the dead, must represent the first stirrings of religious consciousness.

IKEDA: That must be so. Of course, their consciousness was still quite primitive. Nevertheless, I find it extremely interesting that religious awareness seems to have emerged virtually at the same time as human beings themselves.

KIGUCHI: It is said that many animals, sensing their own

death, go to hide somewhere to die alone. For example, one very seldom sees the remains of dead sparrows or doves. It is said that even dogs and cats disappear suddenly when it is time for them to die.

IKEDA: Perhaps this is due to instinct. It is true that, apart from those cases of animals killed by accident, few people have ever witnessed the natural death of wild animals. Even large animals such as elephants apparently contrive to die where their remains will not be openly visible.

SHIMURA: Someone who finds and sells elephant tusks once told me that when an elephant is unfortunate enough to die out in the open, the other elephants of the herd almost always gather tightly around the carcass and carry it away somewhere. If you follow them, you can probably find a large number of tusks and make a profit. However, even docile elephants will attack a human being who takes steps to disrupt the sanctuary of their deceased.

Also I hear that the death of the leader of a flock of eagles is also an impressive spectacle. When he senses the waning of his strength, on a clear day, he soars higher and higher into the sky. Then, from a perfectly cloudless sky, he plunges headlong, shattering his body on the crags of a mountain peak.

IKEDA: I believe I read something like that in the book *Tori no monogatari* (*The Story of Birds*). I recall one passage describing in detail how the eagle flock then silently picks up the scattered pieces of their leader's body, so that nothing will remain.

KIGUCHI: One feels awe at the existence of creatures who, sensing when their end is near, use their death to display noble qualities.

SHIMURA: This must be an instinct inherent in life. However, one feels that people are becoming insensitive to both the dignity inherent in life and the stern reality of death.

KIGUCHI: Generally speaking, people no longer think deeply about life or turn their thoughts toward what is eternal. In place of this we find an overwhelming tendency toward instant gratification and momentary pleasures. This seems to be a unique phenomenon in human history, don't you think?

IKEDA: No, I think this tendency has always been present to some extent. Perhaps we can say, however, that never before has its harmful influence become so thoroughly diffused into every area of society. I feel that our age indeed represents the degenerate Latter Day of the Law described in the Buddhist teachings.

SHIMURA: The trend toward disregard of life seems particularly strong.

KIGUCHI: I don't know if it stems from television or other influences, but recently, the rate of murder and suicide among children and teenagers has been escalating. It's frightening.

SHIMURA: Doing some research, I found some historical texts which indicate that the emperors of the Heian period (794-1185) fully believed that people who do evil acts will fall into hell after they die.

IKEDA: Yes, we find such beliefs expressed, for example, in the poem of Emperor Daigo (reign 897-930) which says:

> I have heard it said –
> When one falls into the depths of hell,
> *Kshatriyas* and *sudras*,
> all are equal.

Kshatriyas were members of the warrior or kingly class in ancient India. Here, we can understand this expression as pointing to the ruling family. *Sudras*, on the other hand, were the untouchables or outcastes.

KIGUCHI: Then this poem can be seen as a reflection of ancient people's belief in a world after death.

SHIMURA: One may easily point out the unsophistication of people of ancient times, but on the whole, modern man makes little effort to come directly to grips with the problem of death. He may fear to think about death. In fact, I believe the fear of death is something that has not changed throughout the ages.

IKEDA: That is certainly true. The writing, *Hōon shō* (*Repaying Debts of Gratitude*) states, 'In these latter times, men become shallow in their wisdom.' Perhaps because they are pressed by daily routines or because they are immersed in a materialistic way of life, their general awareness of death tends to be shallow. We can also say that throughout history, man has not appreciably advanced in his ability to come to terms with the problem of death.

KIGUCHI: Perhaps, because life has become too convenient and people too affluent, the spiritual nature of human beings has become debased.

SHIMURA: The works of the late French author André Malraux continue to captivate many readers. Central to his writing is the search for something eternal, transcending birth and death. He has said this himself, and of course many other people have understood his works in this way.

IKEDA: Malraux's grandfather and father both committed suicide. Moreover, he was taken prisoner by the Nazis during his involvement with the French resistance movement. He no doubt fully witnessed the horror of death.

KIGUCHI: You had a dialogue with Malraux, didn't you? I remember reading its transcription. It impressed me greatly.

IKEDA: That must have been over fifteen years ago. We met twice, in Japan and in France. Certainly Malraux was both an outstanding intellectual and a man of action.

SHIMURA: Malraux sought the eternal in the realm of beauty. I wonder if, in conclusion, he may have reached a resolution in certain respects similar to ideas of Buddhism.

IKEDA: Yes, certainly he did. This is also true of Dr. Toynbee, I believe. Malraux was a real grass-roots diplomat. He's one who should be called a pioneer in international cultural exchange.

SHIMURA: The fact that French cultural treasures, such as the Venus de Milo and the Mona Lisa, were first exhibited abroad in Japan owes much to the efforts of Malraux, who also served as the French Minister of Culture.

The Nine Consciousnesses and the Buddhist Concept of 'Self'

SHIMURA: Anyone who has mastered a particular field of study must eventually be led to look into himself.

KIGUCHI: You mean, led to contemplation of what Buddhism calls the 'self'? It is a vital and extremely complex problem.

IKEDA: That's true. I believe we touched on this earlier, but in clarifying the nature of the 'self', Buddhism expounds the theory of the nine consciousnesses (Skt *vijnāna*).

SHIMURA: We can say that the theory of the nine consciousnesses clarifies the different levels or functions of the mind.

IKEDA: That's correct. However, in terms of Nichiren Daishōnin's Buddhism, which reveals the ultimate Law of life and the universe, I believe we can regard the nine consciousnesses as presenting a view of life in its totality. As I mentioned before, the first five consciousnesses correspond to the sensory functions of sight, hearing, smell, taste and touch. The sixth consciousness can be understood as the power of thought based on intellect and reason. Buddhism further expounds the seventh or *mano*-consciousness, which is a realm extending into the unconscious. *Mano* is a Sanskrit word meaning 'to ponder'. I feel we can understand this seventh consciousness to transcend the realm of everyday thoughts and the relatively shallow discerning power of the sixth consciousness and to include, for example, the kind of abstract thinking involved in academic study or artistic creation.

KIGUCHI: Then can we say that it involves the kind of intense mental efforts one exerts to discover truth or transcendent values, such as the essence of beauty?

IKEDA: Yes, I believe we can. However, when we consider that the seventh consciousness is itself governed by impulses originating from a deeper level of life, and that it is not life's most fundamental dimension, it cannot be called a realm of perfect freedom.

SHIMURA: There is an interesting story in connection with the Austrian psychiatrist Freud. He is famous as the psychiatrist who probed the depths of the human mind. However, one internationally famous German physician said of him that although Freud had discovered the influence of the unconscious mind, he himself was unable to control his own anger, and was subject to fits of convulsive rage.

IKEDA: In this respect, Buddhism turns its attention to the 'self' which generates such drives. In so doing, it clarifies what is called the eighth or *ālaya*-consciousness. The Sanskrit word *ālaya* means dwelling and receptacle. The eighth consciousness is also referred to as the 'storehouse consciousness'. It is so called because an individual's karma is accumulated in this consciousness. The eighth consciousness is also called 'non-vanishing', because, although one's life undergoes the cycle of birth and death, this consciousness does not vanish.

SHIMURA: In other words, the *ālaya*-consciousness continues from one lifetime to the next, carrying with it the whole of an individual's karma.

IKEDA: Yes, that's correct. One's individual life, encompassed by this eighth consciousness, continues even after death, in the state of *kū* or latency. The first five senses and the sixth and seventh consciousnesses, which all function actively while we are alive, recede at the moment of death into a latent state encompassed by the eighth consciousness. That is, the memories, habits and karma stored in the *ālaya*-consciousness moment by moment while one is alive, form the individual 'self' that continues to exist even after death. This consciousness may be thought of as the 'realm' that interweaves all the causes and effects forming one's individual destiny. For example, the 'self' or *ālaya*-consciousness of one who has created for himself the evil karma of the life-tendency of Hell will, in death, 'melt' into the world of Hell inherent in the cosmic life, where it must undergo extreme suffering.

SHIMURA: That is a frightful thought. A black hole must be one such place (laughter).

KIGUCHI: So then, the individual 'self' which fuses with the universal life at death still possesses the life-conditions of the Ten Worlds?

IKEDA: Of course. As the *Sōkanmon shō* states, 'The mutual possession of the Ten Worlds is the true aspect of all phenomenal existences.' Therefore, the individual life that has returned to the universe and persists in the state of latency will continue to experience one or another of the Ten Worlds, depending upon the karma it has created.

KIGUCHI: I believe I mentioned it before, but modern physics confirms that space is itself a realm that manifests various phenomena, according to causality operating from the past. Therefore, although we are, of course, talking about a different dimension, I feel as though I understand the notion of an invisible 'realm' where karma persists.

IKEDA: A 'self' which has established the good karma of a state higher than Humanity will, in death, melt back into the realm of Heaven, Learning, Realization, Bodhisattva or Buddhahood inherent in the cosmic life. For example, a 'self' which has established the state of Heaven will in death enjoy the various pleasures of that state, while one that has established Buddhahood as its basis will experience a joy as vast as the universe itself.

SHIMURA: Can one experience Buddhahood while alive?

IKEDA: Of course – but only through deep and continued faith. However, in discussing how life continues after death, Buddhism places great emphasis on the last moment of one's life as that person's final balancing of accounts.

KIGUCHI: Would you please explain the ninth consciousness?

IKEDA: The *Ongi kuden* states, 'The ninth consciousness corresponds to enlightenment, and the other eight consciousnesses, to delusion.' Therefore, the conclusion of the Buddhist teachings lies in awakening to the ninth consciousness, which can be termed the 'universal self' in the innermost depths of life.

SHIMURA: It is the Buddhism of Nichiren Daishōnin that first defined the true entity of the ninth consciousness.

IKEDA: That's correct. Briefly stated, even the eighth consciousness is subject to the two aspects of purity and impurity and is not firm or unchanging. However, beneath the eighth consciousness, so to speak, lies the original life of the universe – that is, *Nam-myōhō-renge-kyō*, the unchanging reality which reigns over all of life's functions. We carry out our faith because we want to awaken to this fundamental power of life and the cosmos. All those who believe in and practise this Law attain, based on the ninth consciousness, a state of life which enables them to establish a solid and reliable 'self'. Herein, I believe, lies a fundamental and perfect path toward the establishment of a peaceful society.

SHIMURA: The human being is really the key to everything. We must not lose sight of this point.

KIGUCHI: Yes, as that famous passage from the UNESCO charter states, 'Since wars begin in the minds of men, it is in the minds of men that the defences of peace must be constructed.'

IKEDA: It may seem like the long route, but there is no certain road to peace apart from the awakening of individuals. Nichiren Daishōnin's Buddhism teaches that faith makes this possible, although people naturally develop differences in life-condition according to the strength and depth of their faith.

SHIMURA: That's true. Buddhism would lose its meaning if faith did not produce expected results.

The Strict Law of Cause and Effect

IKEDA: The *Sōkanmon shō*, referring to the rebirth of life, states, 'Motivated from within and drawn from without, in the mutual influence of internal and external, harmonizing cause and circumstance ...' As this passage indicates, the karma possessed by the 'self' or individual life in the state of death will, when activated by an appropriate external cause, enable that life to re-emerge and resume its life-activities in this world.

Because the three thousand realms of the external world correspond with those of one's own life, there is not a single exception in the life-state one experiences according to one's own karma. Buddhism is the teaching that clarifies this truth, which accords with the strict law of causality. The theory of causality in itself should actually be discussed in far greater depth, but, for now, let us simply point out that it is inexorable. No human emotions or desires can alter the nature of its workings.

KIGUCHI: Nothing can rival its strict exactitude, which is thoroughly scientific – if I may be permitted to use such an expression.

IKEDA: I think it is quite appropriate. Mr. Toda used to say, 'Science investigates the principles of the natural world, while

Buddhism is a logical religion clarifying the law of causality operating within human life. A true religion is scientific in its attitudes of inquiry; thus there need be no contradiction between religion and science.'

KIGUCHI: It is true that we scientists sometimes become wholly absorbed in investigating the outer world and tend to neglect the inner realm of our own hearts and minds.

IKEDA: The *Sōkanmon shō* states, ' "Cause" means the unified entity of the three truths inherent in all living beings; it is eternal and unchanging.' This passage is a bit difficult, but we can understand 'cause' here to indicate the eternal and unchanging Buddha nature eternally inherent in life throughout the three existences of the past, present and future. In other words, it teaches that all people without exception are endowed with the most respectworthy nature of the Buddha.

SHIMURA: However, even though we are told that we possess the Buddha nature, when we observe the reality of human life, it is sometimes rather difficult to believe.

IKEDA: This question could easily lead to a profound discussion of Buddhist theory. Briefly, the 'external cause' becomes extremely important in this respect. For example, the *Sōkanmon shō* says, 'Even though one is endowed with the three potentials of the Buddha nature, unless he meets a good friend (*zenchishiki*), he can neither awaken to it, nor know it, nor manifest it.' To put it simply, in order to actually manifest our inherent Buddha nature in our daily lives and in our activities in society, we need a 'good friend' as an external cause. From our standpoint, this 'good friend' is the Mystic Law, the teaching which enables us to awaken to our Buddha nature. But because people make no effort to know of this Law, they remain ignorant of it.

The law of cause and effect operating within life is strict indeed. As the *Ongi kuden* states, 'It is strict, and operates through all three thousand realms.' No exception is possible. The law of causality plays no favourites.

KIGUCHI: No compromise is permitted.

IKEDA: The laws and systems of society, no matter how strictly and exhaustively organized, still do not transcend the category of human affairs. They are in no way perfect or absolute – something which is evident from the great divergence in laws and social systems from one country to another.

Birth and Death on the Cosmic Scale

SHIMURA: By the way, Dr. Kiguchi, how much can astronomy tell us about the death of the universe?

KIGUCHI: That's a difficult question. The evolution of the universe since the Big Bang, thought to have occurred some ten to twenty billion years ago, is now the subject of intense research. However, we cannot yet tell, in light of modern science, whether the universe repeats the cycle of birth and death, so to speak, or whether the Big Bang was a once-only occurrence. Moreover, since the death of the universe lies in the distant future, there is no possibility of experimental verification. For this reason, research in this area is not making much headway.

IKEDA: However, I intuitively feel that even if everything in the universe were to come to an end, the universe itself would continue.

SHIMURA: Dr. Kiguchi, would you briefly tell us what forecasts cosmologists have made in this respect?

KIGUCHI: Well, speaking only in terms of the Milky Way, the fixed stars will cease to emit light when their nuclear energy is exhausted. Before that, as they cool, they will expand. If any solar system other than ours exists, some of its planets will be engulfed, just as our own Earth will be. Others will be blown away when their fixed star explodes. Most planets will be drawn away from their stars by the influence of other heavenly bodies.

Other fixed stars will break away from the Galaxy. Tens of billions of years after that, it is said, all matter will melt and 'liquify'. Stars will become black holes, and, according to Stephen Hawking's predictions, eventually evaporate.

IKEDA: So ultimately there will be nothing left.

KIGUCHI: Well, it won't happen for a long, virtually eternal duration of time − that is, not for a number of years approximately equal to ten to the power of anywhere from 10^{26} to 10^{76}.

IKEDA: Buddhism makes references to 'the fire at the end of the kalpa'. This expression refers to the collapse of the physical realm where living beings dwell. In our terms, it would indicate the end of the Earth. As I mentioned before, Buddhism expounds the concept of the four kalpas − the cycle of formation, continuance, decline and disintegration − that all worlds continually undergo. During the Kalpa of Decline, in which everything decays, sentient beings − actually, all living organisms − first become extinct. After that, their world itself is destroyed.

SHIMURA: According to ancient Indian cosmology, in the Kalpa of Decline, the world is consumed in a great fire.

KIGUCHI: To me, that indicates a time when the Earth will be engulfed by the flames of the Sun.

IKEDA: We can find various ways of calculating the lengths of the four kalpas, but in any event, each of them spans a prodigious length of time.

SHIMURA: In that it is a cyclic process, the concept of the four kalpas differs considerably from eschatology, doesn't it? Moreover, it's very clearly defined.

KIGUCHI: I think I mentioned this before, but the death of stars was unknown for a long time. Recently, however, by probing space with infra-red and X-ray telescopes, we are gathering a wealth of new data.

SHIMURA: What is the best-known of such discoveries?

KIGUCHI: Well, for example, the Crab Nebula in the constellation Taurus is quite well-known. We now know that this nebula was originally a massive star that blew itself up.

IKEDA: I've heard that this gaseous nebula was once like our own Sun.

KIGUCHI: That's correct. Moreover, we have discovered the presence, at the centre of that nebula, of a body that emits radio pulses at astonishingly precise, regular intervals. Such bodies are called pulsars. We now know that they are in fact neutron stars as was mentioned before. These stars have been observed to emit extremely powerful bursts of radio energy. The energy of the pulsar derives from its rotation. Its overall rotational energy may not differ greatly from that of the Sun's, but research is being conducted to learn more about how that energy is released so efficiently.

IKEDA: I've heard that its radio energy is more powerful than that of any other heavenly body known to date in our own Milky Way.

KIGUCHI: That's true. It is said that the energy emitted by this pulsar in one second exceeds the Earth's electrical energy requirements for a billion years.

SHIMURA: That's a great deal of energy. Recently I read somewhere that NASA specialists have begun researching the possibility of utilizing this energy.

KIGUCHI: That's probably true. Their nuclear energy represents only 0.8 per cent of their total mass, but if the mass of the star is converted to gravitational potential energy, it could, theoretically, yield one hundred per cent.

IKEDA: The universe is indeed a storehouse of boundless energy. I feel as though it is the mother of all things.

SHIMURA: I've heard that the *Meigetsuki*, the diary of the poet Fujiwara no Teika (1162-1241), contains a reference to the explosion of the Crab Nebula. In the early part of the Shōwa period (beginning in 1926), an amateur Japanese astronomer contributed an article on the diary to an American journal of astronomy as an account of a supernova. This caused quite a stir. It seems that, until that time, we had no historical record of anything that could be precisely identified as a supernova or an exploding star.

KIGUCHI: Even today, many mysteries still surround that nebula. It is the subject of intense research by astronomers the world over.
  A supernova is a star which is exploding. At the other end of the spectrum, so to speak, a star called T-Tauri was also discovered in the constellation Taurus just after it had come into being. Moreover, data concerning the birth of planets was first discovered in conjunction with this constellation.

IKEDA: Then within the same constellation of Taurus, stars are being born and dying simultaneously?

KIGUCHI: Yes, modern astronomy has come to understand that new stars and galaxies are continually being born while others are dying.

SHIMURA: I understand that when stars die and cease to emit light, their temperature drops radically.

KIGUCHI: That's right.

SHIMURA: What is their temperature at that point, roughly speaking?

KIGUCHI: While stars are living, they emit a tremendous amount of energy, and their temperatures reach ten million to thirty million degrees. But when they die, they approach the temperature of space, about three degrees Kelvin.

SHIMURA: What does that equal in degrees centigrade?

KIGUCHI: About minus two hundred and seventy degrees centigrade. This is the mean temperature of space. When a star dies, it cools to the temperature of surrounding space.

IKEDA: In a similar way, when a human being dies, the body naturally no longer generates heat, and gradually begins to cool. However, it will never become cooler than the surrounding temperature.

I've heard that, according to textbooks on forensic medicine, the reason corpses feel cold is because of moisture evaporation from the skin. If the temperature of the air is thirty degrees centigrade, the temperature of the body will also drop to thirty degrees. If the air temperature is ten degrees, the temperature of the body will again drop to that level, but will not become colder.

Even 'Imperishable' Protons Decay

SHIMURA: All stars, living beings and material substances in the universe, including the human body, are composed of elementary particles. I understand that, until just recently, the sub-atomic particles known as protons were generally regarded as indestructible. Protons, as you know, are part of the nuclei of atoms that compose all things. The atomic nucleus is formed of protons and neutrons, and is surrounded by orbiting electrons. Is that correct?

KIGUCHI: Roughly speaking, yes. However, it has become clear that, theoretically at least, even protons eventually decay. They are not orphans excluded from the Buddhist law of impermanence (laughter). Actually, beginning in the summer of 1983, a group in the Department of Science of the University of Tokyo, as well as others in the United States and India, has been involved in experiments researching the decay of protons. They are now trying to confirm their findings experimentally.

SHIMURA: They have set up a laboratory in a mineshaft in the Kamioka mine in Gifu Prefecture.

KIGUCHI: That's right. They've installed an underground tank holding up to three thousand tons of water which is surrounded by rock a thousand metres thick.

IKEDA: These experiments are being conducted on quite a large scale.

KIGUCHI: They have to perform these experiments deep underground in order to isolate them from the influence of cosmic rays.

SHIMURA: In order to confirm the 'death' of protons, they are waiting to record the natural decay of some among the huge number of protons isolated in the water in the tank.

IKEDA: I've heard that, if the universe is considered to be ten billion years old, then the life span of protons is ten billion times longer than that – far beyond human conception.

SHIMURA: Is it possible that the 'death' of such an extremely long-lived and yet tiny invisible particle can be observed?

KIGUCHI: It's going to be quite a challenge. They say that, since the time of the Big Bang until the present, only a small minority of the then existing protons have decayed. However, even though they are extremely long-lived, it stands to reason that among a great number of protons, some will die sooner than others. Therefore, within a year, the death of a few can probably be confirmed. This research into the nature of protons has little immediate relevance to our daily lives. However, through such research, we can better understand how all matter in the universe came to be formed. Thus it should lead to a greater understanding of the birth of all things.

The more science advances, the more we know, but at the same time, the more we realize how little we know. New questions arise, and those enigmas which do remain may prove increasingly complex and difficult to understand. In addition, vast amounts of money and manpower will be needed to unravel these mysteries.

SHIMURA: Incidentally, many of the spin-offs of scientific progress, such as vinyl and polystyrene foam, are undeniably convenient but at the same time present a problem because they do not readily decompose in the natural environment.

KIGUCHI: Birth and death are the way of all things; such is the rhythm of nature. Man-made substances that drastically disrupt this cycle may prove to be ecologically harmful.

SHIMURA: Science has conferred many blessings on people, but occasionally these cause us harm.

IKEDA: Generally speaking, people seek only the blessings of science and remain unaware of its drawbacks. We must remember, however, that science is a two-edged sword.

KIGUCHI: That's quite true. Those who are involved in scientific activities must be keenly aware of this fact. Ours is an age in which science must not run wild; rather, it must be grounded in a philosophy or religion that does not contradict science's basic nature and that can also guide it correctly. This is an absolute necessity.

XIII

Expansion of Man's Perspective of the Universe

> The pursuit of wider views of the universe underlies scientific progress.

Visitors from Outer Space

KIGUCHI: I feel that Mr. Ikeda's recent visits to North and South America[1] demonstrate how Buddhism transcends racial and national boundaries, and can provide all people of the world with the vital energy to renew their lives and achieve happiness.

SHIMURA: Our Buddhist movement does not operate in the dimensions of politics or economics, but in the profound dimension of human life itself. Those ideologically confined to the concept of 'the state' or to differences in political orientation cannot possibly understand its basis.

IKEDA: Science knows no national boundaries. Nor does academic study, or culture. Similarly, I feel, a philosophy or religion, capable of thoroughly convincing people and of providing the means to achieve happiness and satisfaction in life, must also transcend national differences.

KIGUCHI: That stands to reason. For example, the discoveries of Einstein and other scientists, by clarifying principles underlying natural phenomena, have led to a focusing on universal values common to all humanity.

SHIMURA: Moreover, ours is the age in which we must realize that we are all passengers aboard the 'spaceship Earth' and share a common destiny. This becomes all the more apparent when our attention is directed toward outer space.

IKEDA: Incidentally, Peru has an extraordinary feature – the famous 'Nazca drawings', said by some to be evidence of prior contact with extraterrestrials. It is located about an hour away by plane from Lima, the capital city. Huge images are drawn on a dry plateau where the rain almost never falls.

KIGUCHI: Looking out from an airplane flying over the Pan-American highway, one can see the drawings of condors, monkeys, whales and even what might possibly be extraterrestrials.

IKEDA: Because they are so large, on seeing them from the ground, one cannot possibly tell what sort of pictures they are. Only from the air can one perceive their entire image.

SHIMURA: I wonder why they were drawn there.

KIGUCHI: That remains a mystery – which is why people attribute them to visitors from outer space.

IKEDA: According to the report of a scholar in the field of American cultural history, issued about forty-five years ago, they appear to date from somewhere between 300 BC and AD 900. There are several theories concerning their significance and origin, one being that the drawing of these pictures was directed from above by a hot-air balloon. It seems probable that they constitute a chart of the heavens.

SHIMURA: Dr. Kiguchi, I understand it has been formally decided to develop a large-scale rocket in Japan with the same launching power as that of the American space shuttle.

KIGUCHI: Yes, it seems its completion is projected for 1991.

IKEDA: If these plans are realized, will it mean that Japan will then enter the company of 'space nations'?

KIGUCHI: I think so. In addition, I understand there are plans to launch about fifty application satellites in the next fifteen years. We will also take an active part in an American space station project. A Japanese scientist is scheduled to travel on a space shuttle flight four years from now.

IKEDA: He will be the first Japanese in space.

KIGUCHI: That's right. He will be allotted one third of the shuttle, in which he will conduct various experiments in space. However, the cost of his facilities alone will run to about thirty-five million dollars.

IKEDA: That's a considerable expenditure. Many hundreds of thousands of people on Earth are still suffering from poverty and disease. Illiteracy too remains widespread. We must never overlook the misery on Earth in our eagerness to pioneer space.

SHIMURA: More than ten years ago, Mr Ikeda, you suggested the possibility of cooperative international space exploration projects. This seems like a valuable proposal. It would serve both to ease the financial burden on individual nations, and to contribute to the realization of peace, by way of international scientific exchange.

KIGUCHI: The exploration of space will facilitate research in communications, broadcasting, meteorology, metals and even pharmaceuticals in ways not possible on Earth.

IKEDA: Could you give us some examples?

KIGUCHI: Because there is no gravity in space, with special technology, pure chemical compounds can easily be isolated. This could result in the inexpensive large-scale production of specific medications for diabetes, pancreatic disorders and other diseases. It is also said that new alloys and amorphous (non-crystalline) metals will be developed. Space can be thought of as a huge factory with extraordinary capacities.

SHIMURA: The development of space will be a crowning achievement of science. President Ronald Reagan of the United

States, in his New Year's address (January 1984), drew a comparison with the age of the great sea voyages of discovery, and called for the opening of new frontiers of space.

Scientific Progress in the Age of Discovery

IKEDA: About four hundred and sixty years have passed since Magellan made his voyage around the world. The exploits of men such as Columbus and Vasco da Gama also must have served to revolutionize the views of people in those days.

SHIMURA: That's true. For one thing, the truth that the world is round became recognized as an established fact.

KIGUCHI: It's not hard to imagine how men, having finally become liberated from the darkness of the Middle Ages, came to direct their spirit of inquiry toward the unknown: the seas and the heavens.

IKEDA: Thanks to the insatiable curiosity of men who wondered what lay beyond the oceans' horizons, new lands were discovered. I believe that in any age, human beings cherish the great desire to probe and understand the unknown.

SHIMURA: That is why our time is often called the dawn of the space age.

KIGUCHI: It appears that those long sea voyages were made possible, more than anything else, by the development of superior telescopes, maps and compasses.

SHIMURA: Medicine also advanced during that time.

IKEDA: I read somewhere that when Vasco da Gama sailed back into the harbour of Lisbon after completing his expedition that opened the sea route between Europe and India by way of the Cape of Good Hope, only one third of his crew was still alive. Many had died of scurvy due to the lack of vitamin C, and the survivors were suffering from malnutrition as a result of the long sea voyage.

And some of the explorers returned home carrying diseases previously not known in Europe. For that reason, it is said, medical science was compelled to advance.

SHIMURA: When the *Apollo* spaceship returned from its voyage to the Moon, the astronauts were kept in quarantine for several days.

KIGUCHI: Yes, Dr. Sagan and others proposed that this measure be taken.

IKEDA: When I asked Dr. Sagan why, he replied that it was because there existed a possibility of introducing new organisms from space into the Earth's environment.

By the way, a number of scientific advances were made during the age of discovery, especially in Holland.

KIGUCHI: The telescope was invented in Holland, and earlier, the microscope was invented there also.

SHIMURA: Incidentally, the book *Chōsen kagaku gijutsu shi* (*A History of Korean Scientific Technology*) records that 'fire crystals' (lenses) were made in the kingdom of Silla on the Korean Peninsula in the early sixth century.

KIGUCHI: That's about a hundred years before the time of Prince Shōtoku[2] (574-622) in Japan.

SHIMURA: We know from available records that they were what is called black crystal or black quartz.

IKEDA: Among Nichiren Daishōnin's writings, we find the passage, 'When a crystal is faced toward the sun, it produces fire, and when faced toward the moon, it produces water.' It appears that crystals were already being used for various purposes at that time.

SHIMURA: It is said that the microscope was invented by a Dutch optician who happened to superimpose two magnifying glasses, one over the other.

KIGUCHI: And several decades later, another Dutch optician discovered the principle of the telescope quite by chance when he happened to place one pair of eyeglasses in front of another – one for farsightedness and one for nearsightedness.

SHIMURA: Kepler conceived the principle which resulted in our modern visual refracting telescopes. He also invented the famous Keplerian telescope, vastly improving on the one used by Galileo.

IKEDA: So, as the human mind began to decipher the movements of the heavens, man also began to look into the minute constituents of the fabric of life on Earth.

KIGUCHI: Moreover, modern science has come to understand that the vast realm of space and the microcosmic realm of atoms and elementary particles are structurally very closely related.

SHIMURA: I understand that Aristotle held the philosophical opinion that, the more one strives to grasp the principles underlying the workings of the vast universe, the more he will come to understand the nature of the microcosm that is the human being.

IKEDA: In fact, the prefixes *macro-* and *micro-*, respectively denoting large and small, originally derive from the ancient Greek language.

KIGUCHI: Yes, they appear in Aristotle's work, *Natural Philosophy*. Recently the subject of 'optical fibres' has been attracting attention. The technology has developed to the extent that one can operate no less than 5,760 telephone circuits through a glass fibre as slender as a hair.

IKEDA: Something great is embodied in something small, and

something small expands into something great. I feel as though such facts offer another angle from which to comprehend the statement in Nichikan Shōnin's *Sanjū hiden shō* (*The Threefold Secret Teaching*) which reads, 'The entire phenomenal world is encompassed by the one life, and the one life pervades the phenomenal world in its entirety.'

SHIMURA: Even before the age of the great voyages of discovery, people undertook expeditions by land, making, for example, the long trek over the Silk Road to the western regions.

KIGUCHI: Europeans, too, went on arduous journeys seeking the silks and porcelains of the East.

IKEDA: Apparently calendrical studies, the art of printing, gun powder, the compass, mathematics and so forth were originally introduced to the West from the Orient.

KIGUCHI: Was Eastern civilization more highly developed at that time?

SHIMURA: In many respects, yes. This area has been researched by Dr. Toynbee and the British biochemist Joseph Needham, a science historian as well as orientalist at Cambridge.

KIGUCHI: By the fourth century BC, Alexander the Great had already travelled to India.

IKEDA: Among his descendants who remained in the northwestern part of India was one Milinda, also Menander, a Greco-Bactrian king who ruled northern India from his capital at Sāgala in the second century BC. He conducted a famous debate with the Buddhist monk, Nāgasena.

SHIMURA: It is said that King Milinda, who lost the debate, donated a monastery to the Buddhist order and eventually converted to Buddhism.

IKEDA: In the thirteenth century, coming from the opposite direction, Genghis Khan led his expeditionary forces all the way to the eastern gates of Europe.

KIGUCHI: The journey over the Silk Road must have been extremely difficult in those days.

SHIMURA: A part of that journey is recorded in the *Bukkoku ki* (*Record of the Buddhist Kingdoms*) by the Chinese monk Fa-hsien, who journeyed to India around the fifth century in search of Buddhist sutras. I recall a passage which mentions that he used the whitened bones of those who had perished along the way as landmarks.

Moreover, in the seventh century, the Chinese monk Hsüan-tsang embarked on a seventeen-year journey through Central Asia and India. It seems that he was quite prepared to give his life in achieving his purpose.

IKEDA: I cannot forget the words of an archaeologist who once said, 'Perhaps this road still remains today because people risked their lives to establish it. What people build with a life-and-death determination does not easily decay.' Nameless people, bravely risking their lives, crossed the Takla Makan Desert. Forging paths where there had been none, they scaled the rugged T'ien Shan (Tian Shan) Mountains. I feel as though a procession of their spirits must still linger invisibly on that road. Each individual must have devoted his entire lifetime with passion to that journey.

KIGUCHI: That must be true. One can hardly expect to leave any great accomplishments behind with just the incentive to realize gain or profit.

IKEDA: Many of the Egyptian pyramids have suffered great damage. I've heard that the only ones which remain almost intact are the Pyramids of Giza on the outskirts of Cairo, built by numerous people who devoted their lives to the task with a sense of mission and pride.

SHIMURA: That's a keen observation. Incidentally, I've heard that the construction of the pyramids was initiated in part as a kind of measure against unemployment.

IKEDA: Some people think so, but I'm inclined to disagree. I think it is more appropriate to view the pyramids as great enterprises undertaken by people in order to symbolize the formidability of their kingdom.

KIGUCHI: Building the pyramids required great ingenuity. In the room where the pharaoh's body was interred, a type of vent was installed. It was positioned in such a way that, once a year, on a specific day, at a precise moment, the light of a certain star would shine on the pharaoh's casket.
The pyramids are structural wonders from the standpoint of both mathematics and physics. I believe that slave labour alone, undertaken out of a sense of obligation, could never have resulted in such magnificent structures.

Religion for the Space Age

SHIMURA: It appears that a remarkably sophisticated cosmology had developed during the age of the Silk Road.

IKEDA: In fact, it seems entirely possible that Alexander the Great may have been aware of the heliocentric theory.

SHIMURA: The ancient Greek philosophers believed that the Earth revolved around the Sun, didn't they?

IKEDA: Documentary evidence suggests that philosophers such as Aristarchus held a view of the cosmos in which the Sun occupied a central position.

KIGUCHI: Yes, we find something to this effect in a treatise written by Archimedes to Hieron II, the king of Syracuse.

SHIMURA: In that case, what about Genghis Khan? Is it possible that he, too, had a similar view of the universe?

IKEDA: It seems that Greek cosmology was incorporated into Arabian geographical studies and then eventually transmitted to the distant empire of the Mongols. It is said that during the time of Khubilai Khan (grandson of Genghis Khan), a large-scale astronomical observatory was built at Samarkand.

SHIMURA: Then it is possible that Genghis Khan held a view of the cosmos similar to that of the Greeks?

IKEDA: Yes, although we have yet to uncover documents that would positively confirm this.

KIGUCHI: By the way, do the Buddhist teachings suggest that the Earth is round?

IKEDA: Well, to put it another way, the original aim of Buddhism lies in clarifying the essential principle that permeates both human life and the cosmos. In a word, we can say that it reveals the 'Law' as fathomed by the intuitive understanding of the East.

Through that which has form, one approaches the formless. Through formlessness, one approaches eternity. The sutras often speak of the 'major world system' – a view which, while encompassing the Earth, expands to become a concept of the universe as being infinite and of life as being both limitless and eternal. The underlying principle of this view is what we call the Mystic Law.

As to whether the Earth is round, or whether it rotates around the Sun, one should understand that, rather than focusing on such physical dimensions, Buddhism investigates the cosmos from the standpoint of human existence. For that very reason, however, it in no way denies the particular world view which you just mentioned.

Actually, I believe that Buddhism, inheriting the concepts of ancient Indian cosmology, tended to view the Earth as solid and shaped something like a section of a cylinder. However, I recall reading a research report to the effect that during the Kamakura period (1185-1333), one priest among the Daishōnin's disciples used the expression 'the world is round.' I cannot recall the name of the scholar who did the research.

SHIMURA: Would it perhaps have been the late Professor Shinji Maejima (1903-1983) of Keiō University?

IKEDA: Quite probably. He was originally an authority on Islamic history, I understand, but through some coincidental association he came to study the career of Nichiji (b. 1250), one of the Daishōnin's disciples.

SHIMURA: I read that he even reached the Daishōnin's world view.

IKEDA: Nichiji was one of the five senior priests who failed to recognize and maintain the orthodoxy of Nikkō Shōnin's leadership after the Daishōnin's death.

SHIMURA: Fourteen years after Nichiren Daishōnin passed away, Nichiji journeyed to the mainland of China.

IKEDA: That is a well-known story. I find it intriguing that Professor Maejima took an interest in Nichiji.

SHIMURA: According to his own account, during the Second World War, a certain Japanese man happened to find articles that had formerly belonged to Nichiji on sale at a public market-place in a town called Hsüan-hua, not far from the Great Wall of China. He decided to buy them.

However, in the confusion of the evacuation at the end of the war, he was unable to bring these articles back to Japan. Yet he did manage to take photographs of them. Through the assistance of a third party, Dr. Maejima had an opportunity to view the photos. Among them were the photographs of a document written by Nichiji prior to his departure. One passage in it seemed to leap out at him: 'From the eastern parts, I will circle the globe to propagate the teachings.'

IKEDA: So the phrase 'circle the globe' caught his attention.

SHIMURA: Thereupon, Dr. Maejima began to study how Arabian geography had been introduced to China.

KIGUCHI: That makes sense. Arabian geographical studies had taken on the Greek view of the cosmos. If these studies had been transmitted to China, then it stands to reason that they might also have been introduced to Japan.

SHIMURA: Nichiji wrote this document on New Year's Day in the second year of the Einin era – that is, in 1294. However, one does not find the phrase 'circle the globe' in any earlier Japanese source.

IKEDA: Probably just to be sure, Dr. Maejima went so far as to research when globes had first been produced in China. I believe he discovered that they were not made until after Nichiji's use of the phrase 'circle the globe'.

SHIMURA: Twenty-nine years later, to be exact.

KIGUCHI: Amazing. Although the general world view of the Japanese at that time could not possibly have been so advanced, it's conceivable that Nichiren Daishōnin was aware that the Earth is round.

SHIMURA: I was quite astonished myself. Nichiji was forty-six years old when he journeyed to China. Perhaps all that remained of this aging and solitary man was his resolve to 'circle the globe'.

IKEDA: We can't really conjecture about what the Daishōnin's view of the universe was from this single source of information.

Galileo's Conviction

SHIMURA: Discussions about different cosmologies must invariably bring one to the trial of Galileo before the Inquisition.

KIGUCHI: He stated that the geocentric theory, which posits that the Earth is the centre of creation, and the Sun and the other planets orbit around it, appears highly dubious. He said that rather than being a valid theory, it is a mere falsehood.

IKEDA: He is also said to have remarked, 'Do not human beings have eyes? And do they not have a brain? With these, they should investigate even a single point, and if, after thorough investigation, it can no longer be doubted, then they should believe it with all their heart.'

SHIMURA: That's very straightforward. He didn't mince words in the least.

IKEDA: In fact, he seems to have been rather stubborn. Yet he had a strong sense of justice and a tireless seeking spirit for the truth.

SHIMURA: He also seems to have possessed considerable business acumen. However, in his later years he lost sight in both eyes, and his family relationships were unhappy.

KIGUCHI: Galileo also said something to the effect that in this world, while theories are numerous, the things that can be proven are very few. But if one discovers a fundamental theory, then, with that as the basis, many minor postulates can be substantiated.

SHIMURA: It seems that anyone who makes a great discovery or initiates a reform which has some significance has a side to him that most people cannot fathom.

IKEDA: That's why, in the world of the human being, one cannot escape from the jealousy of those who are attached to their own fixed opinions. When someone uncovers a particular truth, then, the more profound that truth, the greater the reaction it will provoke. Its very profundity shines too brilliantly for those who are imprisoned in their old ways of thinking.

KIGUCHI: Rather than acknowledge their error, they instead persecute the person who makes the discovery. This attitude accounts, at least in part, for the actions taken against Galileo. In 1633, Galileo's great work *Dialogue Concerning the Two Chief World Systems: Ptolemaic and Copernican*, which he had written three years earlier, was found to be objectionable. Despite his failing health, he was summoned from Florence to Rome to stand trial before the Inquisition on groundless charges.

SHIMURA: Oddly enough, the *Dialogue* was originally published in 1632 with the full consent of the Vatican censors.

KIGUCHI: Which is why the accusations were completely unfounded. It seems that the Church authorities at first thought that if they simply threatened Galileo, he would immediately retract his position.

SHIMURA: It would appear so. One cardinal told him something to this effect: 'You are of course free to present your theory as a mathematical supposition. But science must be a legitimate and dearly beloved daughter of the Church. To a parent, a daughter is something to be watched over and protected.'

KIGUCHI: Eventually, Galileo went to trial.

SHIMURA: But they didn't give him the death sentence because he was very well-known. By that time, his *Dialogue* had been widely read by educated people.

A Verdict Is Rescinded

KIGUCHI: The trial dragged on for three months. Facing his inquisitors, he said, 'I have tried to prove such absurd propositions as the theory that the Earth orbits around the Sun. What ignorant and careless errors I have made!' And he added,

'I never believed that the Earth moves and that the Sun does not, nor do I now. And I am prepared to prove it.'

SHIMURA: Galileo's testimony on that occasion might appear to have been a compromise with the Catholic doctrine. But the Church has been forced to carry the burden of those words to this very day.

IKEDA: That is quite true. Galileo entrusted himself to the more far-reaching judgment of history. In fact, though placed under house arrest for the next eight years, he devoted himself to research and writing with even more passion and determination than before.

SHIMURA: In 1634, he completed his masterpiece *Dialogue Concerning Two New Sciences*. He also worked out the application of the principle of the pendulum to the regulation of clocks.

IKEDA: There are conflicting opinions, but it would appear that Galileo's famous words, '*E pur si muove*' ('But it [the Earth] does move'), which he supposedly murmured after his recantation, were in fact not actually spoken. However, I believe that in later ages people attributed these words to him as a form of praise, in remembrance of his single-minded resolve to leave behind this important legacy to humanity.

KIGUCHI: Even Descartes, hearing how the trial was developing, hastily concealed in the drawer of his desk a manuscript on the nature of light which he had been writing.

SHIMURA: Among Galileo's students there were some who failed to understand his true intention. And others who feared that they would become involved in the matter.

IKEDA: One finds this sort of thing occurring in all ages and societies. Human beings should fundamentally be both equal and free from any authority.

KIGUCHI: It is just as you say. In 1982, three hundred and fifty years after Galileo's trials, the pope finally admitted, before representative scientists of the world, that the verdict had been an error. I believe that a man demonstrates his courage by admitting fault. As history had demonstrated, a scientific theory is replaced by a new one, and a proven theory may be replaced by another in the future. Herein lies the basis of scientific progress. However, in terms of human happiness and universal peace, I believe that the research of the cosmos should lead to spiritual development. Nevertheless, today, one somehow feels as though only knowledge and information are advancing, while philosophy and religion are at a stalemate. This is why uncertainty prevails to the extent that it does.

IKEDA: Whatever the age, people have held a particular view of the world and the universe, whether lofty or limited, which underlay their progress and development. Moreover, the basis of such views was formed by contemporary philosophy and thought. I believe that in today's space age a new philosophy or religion is necessary to provide a profound conceptual basis for our expanding view of the universe.

NOTES: CHAPTER XIII

1 SGI President Ikeda visited a total of eight cities in three countries (Los Angeles, Dallas, Miami, San Diego and Honolulu in the United States, São Paulo and Brasília, Brazil, and Lima, Peru) between 11 February and 18 March 1984, for peace and cultural exchanges, including meetings with Brazilian President João Baptista de Oliveira Figueiredo, Peruvian President Fernando Belaúnde Terry and other government leaders.
2 Prince Shōtou: regent to the Empress Suiko, he carried out important reforms in government and fostered the development of culture and the spread of Buddhism.

Glossary

Apollo programme A space exploration project carried out in the late 1960s and 1970s by the United States to put men on the Moon.

Aries One of the eighty-eight constellations that can be observed from the Earth, and a sign of the zodiac (the Ram), located between Pisces and Taurus. This star group is best viewed in autumn in the northern hemisphere.

Asōgi (Skt *asamkhya*) 'innumerable': an ancient Indian numerical unit, indicating an uncountably large number. One definition is a number equalling 10^{51}; another is 10^{59}.

Atomic nucleus *See* Nucleus.

Base Any chemical species, ionic or molecular, capable of reacting with an acid to form a salt and water.

Big Bang In cosmology, the theory that the universe began fifteen or twenty billion years ago in a single primeval explosion, which accounts for the universe's continuing state of expansion.

Black hole An object with such tremendous gravitational force that neither light nor matter can escape from it.

Bodhisattvas of the Earth The bodhisattvas entrusted by Shakyamuni with the mission of propagating the Mystic Law in the Latter Day of the Law. They are so called because they appeared from beneath the earth when Shakyamuni Buddha was preaching the *Lotus Sutra*.

Bonten (Skt *Brahma*): One of the two major tutelary gods in Buddhism, the other being Taishaku.

Buddha One who has become awakened to the ultimate truth of life, and who leads others to attain the same enlightenment. In Hinayana Buddhism the word *buddha* is generally used to refer to Shakyamuni. In contrast, Mahayana Buddhism postulates the possible existence of many Buddhas throughout the universe and within the dimension of the eternal flow of time spanning past, present and future. Accordingly, in Mahayana Buddhism the plural, *buddhas*, which include Shakyamuni, is often used.

Carbon tetrachloride A colourless volatile liquid with an odour resembling that of chloroform which is virtually insoluble in water but able to be mixed with many organic liquids. It is used as a solvent and dry-cleaning agent.

Chih-i *See* T'ien-t'ai.

Clone An organism genetically identical to its parent which is produced by some form of non-sexual reproduction or by an egg developing without being fertilized by a sperm.

Comet A small celestial body that travels around the Sun in an elliptical orbit.

Conservation of charge The principle that the total net charge in an isolated system remains constant.

Conservation of energy The principle in classical physics that the total energy in a system remains constant. Energy can neither be created nor destroyed, although it can be changed from one form to another.

Conservation of momentum Also known as momentum conservation. The principle which states that the total linear momentum of an isolated system remains constant in magnitude and direction when no external force is applied to the system.

Corona Also called solar corona. The Sun's upper atmosphere which becomes visible during a total solar eclipse.

Cosmic dust Fine particles of solid matter present throughout interstellar space.

Cosmic rays Also called cosmic radiation. Electrons and atomic nuclei from outer space that strike the Earth at almost the speed of light.

Crab Nebula A gaseous nebula in the constellation Taurus, five thousand light-years from the Earth. It is thought to be the remnant of a supernova recorded by Chinese astronomers in 1054.

Cro-Magnon man A tall, erect race of men who inhabited southern Europe and are considered to be the recent predecessor of modern man; so named because the first fossil remains were found in Cromagnon cave near Les Eyzies, Dordogne, France, in 1868.

Daimoku The invocation of *Nam-myōhō-renge-kyō*, in Nichiren Daishōnin's Buddhism.

Daishōnin Literally, 'great sage'. An honorific title for Nichiren, founder of the Nichiren Shōshū sect of Buddhism.

Dark nebulae Clouds of particles in the Milky Way which cannot be seen unless silhouetted against bright nebulae or rich star fields.

Dengyō (767-822): the founder of the Tendai sect in Japan, who asserted the supremacy of the *Lotus Sutra*.

Diffraction The spreading and bending of waves as they pass through an aperture or around the edge of an object.

$E = mc^2$ *See* Einstein's (mass-energy) equation.

Eagle Peak (Skt *Gridhrakūta*): sometimes called Vulture Peak. A mountain located to the northeast of Rājagriha, the capital of Magadha in ancient India, where Shakyamuni is said to have expounded the *Lotus Sutra* and other teachings.

Eighty thousand teachings Also eighty-four thousand teachings. All the teachings which Shakyamuni Buddha expounded during his lifetime. The figure 'eighty thousand' is not intended to be literal but simply indicates a large number.

Einstein's (mass-energy) equation The relationship derived by Albert Einstein in 1905 ($E = mC^2$) which states that the energy of a system is equivalent to its mass times the square of the speed of light.

Electro-magnetic interaction An attraction or repulsive force that functions between charged particles. It controls atomic structure, chemical reactions and all electro-magnetic phenomena.

Elementary particles Also called sub-atomic particles. The fundamental units of matter and energy, e.g. protons, neutrons, electrons, mesons, etc.

Encke's Comet A very small, faint comet with a period of 3.3 years, the shortest period of any known comet.

Essential teaching The latter half of the twenty-eight-chapter *Lotus Sutra*, which is Kumārajīva's Chinese translation. In the section on the 'essential teaching', Shakyamuni casts off his transient role as the historical Shakyamuni who first attained enlightenment during his lifetime in India, and he asserts that he actually attained enlightenment in the immeasurably distant past called *gohyaku-jintengō*.

Evolution A theory postulated in the nineteenth century which explains that the existence of the present forms of plant and animal life resulted from a gradual evolutionary process which began with the earliest and most primitive organisms. Darwin's theory of natural selection stood in marked contrast to the notion of divine creation.

Five components of life Also five *skandhas*. Form, perception, conception, volition and consciousness. These five components unite temporarily to form an individual living being. Form is the physical aspect of life and includes the five sense organs – eyes, ears, nose, tongue and body – through which one perceives the external world. Perception is the function of receiving external information through the six sense organs (the five sense organs plus the 'mind', which integrates the impressions of the five senses). Conception is the function of creating mental ideas or conceptions about what has been perceived. Volition is the will which acts on the conception and motivates action toward what has been perceived. Consciousness is the function of discernment, and gives rise to the components of perception, conception and volition.

Five signs of decay Also five types of decay. Five signs of decline which appear when the life of a heavenly being is about to end. They are: (1) his clothes become soiled, (2) the flowers on his head wither, (3) his body smells bad and becomes dirty, (4) he sweats under the armpits, and (5) he does not feel happy wherever he may be. These signs indicate the impermanence and transience of joy and rapture in Heaven.

Four continents The continents situated respectively to the east, west, north and south of Mount Sumeru, according to the ancient Indian world view. One of them, Jambudvīpa, is said to correspond to our world. The expression 'four continents' is also used to refer to this world.

Four evil paths The life-conditions of Hell, Hunger, Animality and Anger, the lowest four of the Ten Worlds.

Four Heavenly Kings The gods said to live halfway down the four sides of Mount Sumeru. They serve the god Taishaku living on the top of Mount Sumeru as his generals and, in Buddhism, are regarded as protective deities.

Four kalpas Four periods of time corresponding to the four stages in the cycle of formation, continuance, decline and disintegration which a world is said to repeatedly undergo. In the Kalpa of Formation, a world takes shape in space, and a variety of sentient beings appear on it. In the Kalpa of Continuance, living beings conduct their life-activities. In the Kalpa of Decline, the land is destroyed by natural disasters, and living beings gradually diminish and then disappear completely. In the Kalpa of Disintegration, complete destruction has taken place, and everything has entered the phase of non-existence. According to the Buddhist view, a world and the universe undergo this cycle repeatedly throughout eternity.

Gamma-rays Spontaneously emitted electro-magnetic radiation arising from radioactive substances in the process of a nuclear transition.

General theory of relativity *See* Theory of relativity.

Genetic code The genetic information controlling the inheritance of characteristics from one generation to the next. It is expressed by a linear sequence of nucleotides in the DNA of chromosomes.

God of the moon The deification of the Moon. This god was adopted by Buddhist sects as a protective deity.

God of the stars The deification of the stars in Buddhism. The expression 'god of the stars' is also used to refer specifically to the deification of Venus.

God of the sun The deification of the Sun. This god was adopted by Buddhist sects as a protective deity.

Gohonzon The object of worship of the Nichiren Shōshū sect. This sect expounds that the Gohonzon is the embodiment of the ultimate Law permeating life and the universe.

Gohyaku-jintengō An incredibly long period of time described in the *Juryō* chapter of the *Lotus Sutra* which is far longer than *sanzen-jintengō*, a period of time described in the same sutra. *Gohyaku-jintengō* indicates how much time has elapsed since

Shakyamuni's original enlightenment.

Gongyō The recitation of Buddhist sutras in front of an object of worship, though the exact ritual and method of practice differ according to the sect of Buddhism. In Nichiren Shōshū, *gongyō* means to chant *Nam-myōhō-renge-kyō* and recite part of the *Hōben* chapter and the entire *Juryō* chapter of the *Lotus Sutra* in front of the Gohonzon.

Grand unified field theory Any theory which attempts to express the four known physical interactions (i.e. gravitational, electro-magnetic, weak and strong interactions) within a single framework. The verification of a theory which relates three interactions, excluding gravity, is expected. This theory is also commonly referred to as the grand unified field theory.

Gravitational waves A propagating gravitational field travelling at the speed of light which is produced by some change in the distribution of matter. It exerts a force on masses in its path.

Great citadel of the hell of incessant suffering *See* Hell of incessant suffering.

Halley's Comet A bright comet which appears approximately every seventy-six years, named after Edmund Halley (1656-1742) who first calculated its orbit in 1705. Last seen in 1910, it will be visible in the early spring of 1986.

Hell of incessant suffering (Skt *Avīchi*): the most terrible of the eight hot hells described in the Buddhist scriptures. It is so called because those who inhabit it are said to suffer without a moment's respite. This hell is also called the great citadel of the hell of incessant suffering because it is very difficult for its inhabitants to escape from it.

Hinayana Buddhism One of the two major streams of Buddhism, the other being Mahayana. Hinayana teachings regard earthly desires as the cause of suffering, and assert that suffering is eliminated only by eradicating earthly desires. Followers of the Hinayana teachings ultimately aim at attaining the state of *arhat*.

Hokke sect 'Lotus sect': originally another name of the Tendai sect, which was so called because it bases itself on the *Lotus Sutra*. After the advent of Nichiren Daishōnin, who also asserted the supremacy of the *Lotus Sutra* among all of Shakyamuni's teachings, the term *Hokke sect* is also used to refer to the Buddhism which he established.

Hours of the Ox and the Tiger From 2.00 am to 4.00 am. The twelve signs of the Oriental zodiac cycle were used to indicate time of day as well as succession of years. Introduced from China, this cycle includes the names of nine mammals, one bird and two reptiles.

Hubble constant The velocity at which the galaxies recede with increasing distance.

Ichinen Literally, 'one mind'. Also translated as 'life-moment' or 'life-essence'. *Ichinen* or life-essence manifests itself at each moment. *Ichinen* also means life's true nature or ultimate reality.

Ichinen sanzen 'A single life-moment possesses three thousand realms.' A philosophical system set forth by T'ien-t'ai. *Ichinen* (one mind, life-moment or life-essence) is the life that continues to exist for eternity, and *sanzen* (three thousand), the varying aspects and phases it assumes at each moment. The expression 'three thousand' is an integration of the Ten Worlds, their mutual possession, the ten factors and the three realms of existence. Multiplied ($10 \times 10 \times 10 \times 3$), these figures yield three thousand.

Immutable karma One of two types of karma, the other being mutable karma. Immutable karma is karma which inevitably gives rise to a fixed result. Immutable karma was traditionally considered unchangeable. Nichiren Daishōnin asserted that faith in the Mystic Law will eradicate even immutable karma, to say nothing of karma which is mutable. The life span of an individual and the circumstances of his death are included within the category of immutable karma.

Indeterminacy principle Also known as the uncertainty principle. The idea that it is impossible to accurately measure both the position and the velocity of an electron. In a broader sense, the precept states that the act of measuring something will invariably affect the thing being measured and that a precise determination is therefore impossible.

Infra-red Astronomical Satellite (IRAS) A tri-partite project by the United States, Great Britain and the Netherlands to detect weak infra-red rays emanating from distant parts of the cosmos.

Infra-red telescope: An astronomical instrument that converts an invisible infra-red image into a visible image and enlarges this image much like an optical telescope.

Interstellar matter The gaseous and dust material between the stars.

Ion An atom, molecule, or group of atoms or molecules that carry a positive or negative charge as a result of having gained or lost one or more electrons.

IRAS-Araki-Alcock Comet A new comet discovered on 3 May 1983. The comet is named after those who discovered it.

Iron bacteria Bacteria which act on iron in such a way as to oxidize ferrous iron to the ferric state.

Jōsei Toda (1900-1958): the second president of a lay organization of Nichiren Shōshū Buddhism called Sōka Gakkai, which follows and propagates the teachings of Nichiren Daishōnin. In 1943, because of their opposition to Japanese militarism, which opposed religious freedom, Toda and Tsunesaburō Makiguchi, the Sōka Gakkai's first president, were arrested and imprisoned. In 1944, one year before the Second World War drew to a close, Makiguchi died in prison. Thereafter, Toda was released on parole, and he began his efforts to reorganize the society and dedicated his remaining years to the propagation of Nichiren Daishōnin's Buddhism.

Karma Potential effects residing in the inner realm of life which manifest themselves as various results in the future. Buddhism teaches that every action, both good and evil, imprints a latent influence in one's life. According to this concept, one's actions in the past have shaped his reality at present, and his actions in the present in turn determine his future. This law of karmic causality is said to operate over the three existences of past, present and future, and it is karma formed in past lifetimes which accounts for the differences with which we are born into this world.

Kepler's laws Laws formulated by Johannes Kepler that describe the motions of planets in their orbits.

Kōsen-rufu Literally, to 'widely declare and spread (Buddhism)'. Nichiren Shōshū and its lay organization, the Sōka Gakkai, are aiming to achieve *kōsen-rufu* or the worldwide propagation of Buddhism with the belief that it enables each individual to realize true happiness, and results in the establishment of lasting peace on Earth.

Kū (Skt *shūnya* or *shunyatā*): a fundamental Buddhist concept, variously translated as non-substantiality, emptiness, void, latency, relativity, etc. The concept that entities do not have a fixed or independent existence. *Kū* also means a potential state that is considered neither existence nor non-existence.

Kuon ganjo Time without beginning. The term *kuon ganjo* indicates that dimension which is outside the temporal framework, having neither beginning nor end.

Latter Day of the Law The last of the three periods following Shakyamuni's death. It is predicted to last for more than ten thousand years. During this period Buddhism falls into confusion and Shakyamuni's teachings lose their power to lead people to their enlightenment. However, it is also the time when the essence of the *Lotus Sutra* is revealed and propagated in order to enable people to realize Buddhahood. The term 'Latter Day of the Law' is often used to indicate the time period which is the present era.

Law of universal gravitation Originally proposed by Newton in 1687 to explain the motion of the planets, it states that particles of matter attract each other in inverse proportion to the square of the distance between them and in direct proportion to the product of their masses.

Lexell's Comet Seen only one time in 1770, this small comet approached to within 3.2 million kilometres of the Earth.

Life-moment *See* Ichinen.

Lotus Sutra (Skt *Saddharma-pundarīka-sūtra*): the sutra which explains how the potential for enlightenment exists within all living beings and reveals that Shakyamuni originally attained enlightenment in the inconceivably distant past. Among the three extant Chinese translations, the *Myōhō-renge-kyō* in twenty-eight chapters, by Kumārajīva in 406, exerted the greatest influence and was the most widely read. In China and Japan, therefore, reference to the *Lotus Sutra* usually indicates the *Myōhō-renge-kyō*.

Mahayana Buddhism One of the two major streams of Buddhism, the other being Hinayana. Mayahana teachings emphasize practice amid the realities of society for the benefit of all people. Mahayana expounds that the bodhisattva practice is the means toward realizing the enlightenment of both oneself and of others.

Major world system One of the world systems in ancient Indian cosmology. One major world system comprises one billion worlds. Here one world is similar to today's concept of a solar system. There were thought to be countless major world systems in the universe.

Mandala An object of worship in which Buddhas and bodhisattvas are depicted or in which a mystic doctrine is written.

Mantle The intermediate zone of the Earth surrounding the core and surrounded by the crust.

Mass A property of matter that constitutes one of the fundamental, undefined quantities upon which all physical measurements are based. Mass is a measure of a body's inertia or its resistance to acceleration.

Meteor Phenomena which are seen when a body from space (a meteoroid) passes through the Earth's atmosphere. This includes a flash and a streak of light as well as an ionized trail.

Meteorite A meteor which reaches the Earth's surface without being vaporized.

Meteoroid Any solid object moving in interplanetary space that is smaller than a planet or asteroid but larger than a molecule.

Middle Way The true nature of all things which neither is born nor dies, and which cannot be defined by either of the two extremes, existence or non-existence, but manifests characteristics of both. The term Middle Way is also used to refer to the ultimate truth of all phenomena.

Mutual possession of the Ten Worlds A principle that each of the Ten Worlds, or ten life-conditions, potentially contains all ten within itself. One of the component principles of *ichinen sanzen*. The significance of this principle is that all human beings have the potential for Buddhahood.

Myōhō-renge-kyō (1) A Chinese translation of Shakyamuni's *Lotus Sutra*, made by Kumārajīva. It consists of eight volumes and twenty-eight chapters. (2) The term *Myōhō-*

renge-kyō means the entity of the Mystic Law itself, or *Nam-myōhō-renge-kyō*.

Mystic Law The ultimate Law of life and the universe. In Nichiren Daishōnin's Buddhism this term is used to indicate the Law of *Nam-myōhō-renge-kyō*.

Nam-myōhō-renge-kyō The ultimate Law or true entity of life permeating all phenomena in the universe which was revealed by Nichiren Daishōnin.

Nayuta (Skt) An Indian numerical unit. The *Abhidharma-kosha-shāstra* (Jap *Kusha ron*) defines it as one hundred billion (10^{11}), though there are other interpretations.

Neanderthal man A race of prehistoric men who inhabited much of Europe and the area surrounding the Mediterranean between two million and ten thousand years ago. Named after the fossil remains which were first discovered in 1856 in the Neanderthal Valley near Düsseldorf in what is presently West Germany.

Nebular hypothesis The theory of the origin of the solar system originally proposed by Immanuel Kant (1755) and later refined by Pierre-Simon Laplace (1796). It assumes that the solar system evolved from a hot gaseous nebula.

Neutrino A stable uncharged elementary particle with zero mass.

Neutron An elementary particle having approximately the same mass as a proton but lacking an electric charge.

Neutron bomb Also known as a neutron radiation weapon. A nuclear weapon with a small yield (1 – 2 kilotons) which causes minimal blast damage but emits lethal neutron and gamma radiation in a relatively limited area (radius 1 – 2 kilometres).

Neutron star A high-density star thought to be in its final stage of stellar evolution consisting mainly of neutrons. Neutron stars are thought to be high-energy radio sources known as pulsars. When neutron stars further contract and increase in density, they are thought to become black holes. *See also* Black hole.

Newtonian mechanics The system of classical mechanics based upon the laws of motion formulated by Newton. Mass and energy are treated as separate, conservable, mechanical properties.

Nichiren Daishōnin (1222-1282): the founder of what is now known as Nichiren Shōshū, which regards him as the Buddha of the Latter Day of the Law. He clarified that the ultimate Law permeating life and all phenomena in the universe is the essence of the *Lotus Sutra*, the Mystic Truth or *Nam-myōhō-renge-kyō*.

Nichiren Shōshū A sect of Buddhism which regards Nichiren Daishōnin as its founder and Nikkō as his true successor. Its head temple, Taiseki-ji, is in Japan. The sect aims at enabling all humankind to attain absolute happiness by propagating the Law of *Nam-myōho-renge-kyō*.

Ninth consciousness The core of all spiritual functions which is at the innermost depths of life. The last of the nine consciousnesses, or nine kinds of discernments – which incorporate both the faculties of conscious discernment and those below the level of consciousness – is independent of all karmic impurities and is identified with the true entity of life.

Nova A faint variable star which suddenly increases in brightness, sometimes reaching a mean magnitude of -14. Novae have been incorrectly called 'new stars' because of their sudden appearance.

Nuclear force Also called strong interaction. In nuclear physics, one of the four different interactions which account for all forces observed in the universe. This is a very strong force, which acts only at the short distances between nucleons and is responsible for holding the atomic nucleus together.

Nucleon A general name for a proton or neutron, the building blocks of atomic nuclei.

Nucleus The most massive part of the atom which forms its central core. Nuclei are positively charged and consist of protons and neutrons, collectively called nucleons.

One mind *See* Ichinen.

Ongi kuden 'Record of the Orally Transmitted Teachings.' A compilation of Nichiren Daishōnin's lectures on the *Lotus Sutra*.

Opposition The position in which a planet, with its orbit outside that of the Earth, is in alignment with the Earth between it and the Sun. A planet in opposition can be easily observed because at that point it is closer to the Earth.

Ozone A bluish gas, O_3, produced in the stratosphere when high-energy ultra-violet radiation strikes oxygen molecules.

The ozone thus formed acts as a screen for ultra-violet radiation.

Paleozoic The era of geological time occurring between two hundred and twenty million and six hundred million years ago, characterized by the appearance of all classes of invertebrates (except insects). Fish, reptiles and amphibians appeared toward the end of this era.

Perihelion The point in the solar orbit of a planet, comet or artificial satellite at which it is nearest to the Sun.

Photo-dissociation The absorption of a quantum of electromagnetic energy which results in the removal of one or more atoms from a molecule.

Photosynthesis The chemical process by which green plants release oxygen and synthesize organic compounds from carbon dioxide and water in the presence of sunlight.

Proton A positively-charged elementary particle and a building block of all atomic nuclei.

Proto-planet A condensation that gradually formed into a planet in the gas or dust clouds from which the planetary system is assumed to have originated.

Proto-star A disc-shaped or flattened mass of gas that is thought to condense and develop into a star.

Provisional and true teachings Classification of Shakyamuni's teachings. The provisional teachings are those which Shakyamuni expounded as a temporary means to lead people to the true teaching. The provisional teachings reveal only partial aspects of the truth which the Buddha realized, while the true teaching expounds the truth in its entirety.

Pugwash Conferences First held in Pugwash, Nova Scotia, in 1957, in which leading scholars and scientists met to discuss ways to stop the arms race and reduce nuclear armaments. The social responsibility of scientists has continued to be discussed at these conferences with regard to the sciences and world affairs.

Pulsar A cosmic source of short and intense radio emissions which appears to pulsate.

Quantum mechanics Also known as the quantum theory. A modern theory concerning the mechanics of atoms, sub-atomic particles, molecules and electro-magnetic radiation which could not be explained with classical physics.

Radio-telescope Actually not a telescope, but the combination of a radio receiver and a highly-directional antenna used to measure the amount of radio energy coming from various directions in the cosmos.

Radio-isotope An isotope which exhibits radioactivity, often used for radio labelling. Isotopes of an element have the same atomic number but a different atomic mass.

Red giant The evolution of a star, which has progressed to the extent that the burning of the hydrogen core has been completed, the helium core has become denser and hotter, and the envelope has expanded to about one hundred times its initial size.

Reflecting telescope Also known as a reflector telescope. A telescope in which a concave parabolic mirror gathers light and forms a real image of an object.

Russell-Einstein statement A manifesto issued in July 1955 in London, drafted by Bertrand Russell, outlining the risks of nuclear war. It was signed by Albert Einstein two days before he died and was an important step toward the establishment of the Pugwash Conferences. *See* Pugwash Conferences.

Sahā world This world where people live. The Sanskrit word *sahā* means endurance. The *sahā* world is so called because people in this world must undergo many sufferings.

Sanzen-jintengō An immensely long period of time described in the *Kejōyu* chapter of the *Lotus Sutra* to indicate the amount of time which has passed since Shakyamuni preached the *Lotus Sutra* to his disciples in an existence as the sixteenth son of the Buddha called Daitsū.

'Self' The essence of life as grasped by Buddhism. According to the Mahayana concept of *kū* or non-substantiality, a living entity does not have a fixed or independent existence. Therefore, the 'self' as expounded in Buddhism is often used to refer to the true nature of life, in other words, to the Law permeating a living entity and the universe.

Shakubuku: A method of propagating Buddhism by refuting another's attachment to heretical views and thus leading him to the correct Buddhist teaching.

Shōju A method of propagating Buddhism by gradually leading a person to the truth without refuting his attachment to a lower teaching. The term is used in contrast to *shakubuku*, or propagation by strongly refuting another's mistaken views.

Simultaneity of cause and effect The principle that cause and effect exist simultaneously in a single life-moment. Here 'cause' refers to the nine worlds of delusion in which the Buddha nature still remains dormant, and 'effect', to Buddhahood or enlightenment. Therefore, in terms of practice, this concept means that the latent Buddha nature (effect) emerges from within people's lives (cause).

Six paths The life-conditions of Hell, Hunger, Animality, Anger, Humanity (or Tranquillity) and Heaven (Rapture), the first six of the Ten Worlds which one experiences at each moment. The six paths indicate states of delusion or suffering. One who is in these states is governed by his reactions to external stimuli and is, therefore, never really free but always at the mercy of changing circumstances.

Solar wind A continuous flow of gases such as ionized hydrogen and helium from the Sun, throughout the solar system.

Sōka Gakkai A religious society which was founded in 1930 as a lay organization of Nichiren Shōshū Buddhists, which aims at propagating Nichiren Daishōnin's teachings and thereby securing the happiness of each individual and establishing world peace.

Special theory of relativity *See* Theory of relativity.

Spectral type The classification of stars on the basis of their spectrum from which various factors concerning their physical and chemical structure and surface temperatures can be determined.

Sumeru A mountain thought to be at the centre of the world, according to ancient Indian tradition.

Sunspot A dark area on the Sun's surface caused by a decrease in the surface temperatures in a particular location. The number of sunspots changes during an eleven-year cycle.

Supernova A star that suddenly bursts into great brilliance after it has exploded; its magnitude is brighter than that of a nova. *See also* Nova.

Taishaku (Skt *Indra*): one of the two main tutelary gods in Buddhism, the other being Bonten. He lives in the

Trāyastrimsha Heaven on the summit of Mount Sumeru. While Shakyamuni was engaged in bodhisattva practice, Taishaku is said to have assumed various forms to test his resolve.

Tathāgata A title of the Buddha.

Taurus: One of the eighty-eight constellations that can be observed from the Earth, and is also a sign of the zodiac (the Bull), located between Aries and Gemini. This star group is best viewed in winter in the northern hemisphere.

Ten directions The entire dimension of space: north, south, east, west, northwest, northeast, southeast, southwest, up and down.

Ten non-dualities Also ten onenesses. Ten principles set forth by Miao-lo (711-782), the sixth patriarch in the lineage of the T'ien-t'ai school in China. This concept clarifies the inseparability of phenomena and living beings. The ten non-dualities of oneness include: the oneness of body and mind, the oneness of cause and effect, the oneness of life and its environment, and so forth.

Ten Worlds Ten life-conditions which a single entity of life manifests. The Ten Worlds is a component principle of *ichinen sanzen*. They are: (1) The state of Hell, or the condition in which one is dominated by the impulse, in a fit of rage, to destroy himself and everything else. One is utterly devoid of freedom and undergoes extreme and indescribable suffering. (2) The state of Hunger, or the condition characterized by the insatiable desire for food, clothes, wealth, pleasure, fame, power and so forth. In this state, one is tormented by relentless cravings and by his inability to assuage them. (3) The state of Animality, or the condition governed by instinct, in which one has no sense of reason or morality. In the state of Animality, one stands in fear of the strong but despises and preys upon those weaker than himself. (4) The state of Anger, or the condition dominated by a selfish ego. In this state, one is compelled by the need to feel superior to other people in all matters, despising them and valuing his life only. (5) The state of Humanity or Tranquillity, or the condition in which one can make fair judgments, control his instinctive desires through reason and act in harmony with his environment. (6) The state of Heaven or Rapture, or the condition in which one feels great pleasure at having his desires fulfilled. (7) The state of Learning, or the condition in which one awakens to the impermanence of all things and the instability of the lower six states, and seeks lasting truth, aiming at self-reformation through the teachings of others. (8) The state of Realization, or the condition in which one perceives the impermanence of all phenomena and strives to free himself from the sufferings of the lower six states from Hell through Heaven by seeking some lasting truth through his own observations and effort. (9) The state of Bodhisattva, or the condition in which one not only aspires for enlightenment himself but also devotes himself to compassionate actions for the sake of other people. (10) The state of Buddhahood, or the condition of perfect and absolute freedom, in which one enjoys boundless wisdom and compassion, and is filled with the courage and power to surmount all hardships.

Theoretical teaching The first half of the twenty-eight-chapter *Lotus Sutra*, which is Kumārajīva's Chinese translation. The theoretical teaching takes the form of preaching by a provisional Buddha, the historical Shakyamuni Buddha who is depicted as having first attained enlightenment during his lifetime in India. The 'theoretical teaching' expounds that all people are potential Buddhas.

Theory of atoms The theory proposed by the Greek philosophers Leucippus and Democritus in the fifth century BC that speculated that all matter can be accounted for by combinations of innumerable hard, small, indivisible particles (called atoms) of various sizes but of the same basic material.

Theory of relativity A theory of mechanics developed by Einstein to account for the discrepancies in Newtonian mechanics with regard to high-speed relative motion. The special theory refers to inertial, that is to say, non-accelerated, frames of reference. The theory states that the speed of light is constant throughout the universe and is independent of the speed of the observer. A conclusion of the theory is that the mass of a body is a measure of its energy content, as embodied in the equation $E = mc^2$. The general theory, an extension of Einstein's earlier theory to include accelerated systems, makes possible an analysis of gravitation.

Theory of the four elements A theory proposed by Greek philosophers and espoused by Aristotle which explains the composition of the universe and its origin. Earth, Water, Air and Fire were believed to be different forms of the same substance, occupying their own spheres in the universe. Combined in varying proportions, they accounted for the apparent differences in the nature and quality of the physical world.

Three bodies of the Buddha Also called the three properties or three enlightened properties. Three kinds of body which a Buddha may possess. They are: (1) The Dharma body. The fundamental truth or Law to which the Buddha is enlightened. (2) The bliss body, which is obtained as the reward of completing bodhisattva practices and having understood the supreme wisdom. (3) The manifested body, or the physical form in which the Buddha appears in this world in order to save the people. There are two views of this concept. One is that the three bodies exist separately, while the other view regards these three as integeral within the life of one Buddha.

Three evil paths The life-conditions of Hell, Hunger and Animality, the lowest three of the Ten Worlds. The term 'three evil paths' is also used to indicate the realms of suffering into which one falls as a result of having committed evil deeds.

Three existences Past, present and future. The dimension of time. According to Buddhism, the three aspects of the eternity of life are fused into unison by the law of the simultaneity of cause and effect.

Threefold world The world of desire, the world of form and the world of formlessness. The realms inhabited by unenlightened beings who transmigrate within the six paths. The world of desire includes our human world. The world of desire is so called because its inhabitants are ruled by various desires. The world of form, located above the world of desire, is the realm in which its inhabitants have material form but are free from desires. The world of formlessness, located above the world of form, is free from both desire and from the restrictions of matter. The term 'threefold world' is often used to indicate this human world.

Three obstacles and four devils A categorization of the various obstacles and hindrances which interfere with one's practice of Buddhism.

Three poisons Greed, anger and stupidity, or ignorance. The fundamental evils inherent in life which give rise to human suffering. The three poisons are often regarded as the source of all illusions and desires and are so called because they defile and torment people's lives.

Three potentials of the Buddha nature The three essentials which work together as causes to enable one to attain Buddhahood. They are: innate Buddhahood; the wisdom to perceive it; and good causes, or Buddhist practice, to develop this wisdom. The third of the three potentials of the Buddha nature is considered a potential because everyone is endowed with the innate will and power to motivate themselves into action.

Three truths The truth of non-substantiality (Jap *kūtai*), the truth of temporary existence (*ketai*) and the truth of the Middle Way (*chūtai*). Three integral aspects of the truth formulated by T'ien-t'ai. The truth of non-substantiality means that phenomena have no independent or fixed existence of their own; their true nature is *kū*, the potential state that is neither existence nor non-existence. The truth of temporary existence means that while all things are *kū* or non-substantial in nature, they nevertheless possess a provisional or temporary reality which undergoes constant change. The truth of the Middle Way is that all phenomena are characterized by both non-substantiality and temporary existence yet are in essence neither.

Three treasures The three things which all Buddhists should revere and serve. They are the Buddha, the Law (Skt Dharma) and the Priesthood (Samgha).

T'ien-t'ai (538-597): also called Chih-i. The founder of the Chinese T'ien-t'ai school, who expounded the theory of *ichinen sanzen* on the basis of the *Lotus Sutra*.

Triangulation A method for measuring the linear distance to a remote point or for finding one's position on Earth or at sea.

Twelve-linked chain of dependent origination Also twelve-linked chain of causation, etc. An early teaching of Buddhism depicting the causal relationship between a person's ignorance and his sufferings.

Two vehicles The teachings expounded for people in the conditions of Learning (Skt *shrāvaka*, Jap *shōmon*) and Realization (*pratyekabuddha, engaku*). The vehicle of Learning leads one to the state of *arhat* through the teachings of the four noble truths (Skt *catur-ārya-satya*), and the vehicle of Realization leads one to the state of *pratyekabuddha* through the teaching of the twelve-linked chain of causation (*dvādasha-anga*).

Ultra-violet radiation Electro-magnetic radiation falling between violet light and the longer X-ray portions of the electro-magnetic spectrum.

Unified field theory A theory which attempts to explain electro-magnetic and weak interactions within a single unified framework.

Variable star A star that exhibits a detectable change in brightness. The change in brightness can be periodic like clockwork or completely erratic.

Weak interaction A kind of fundamental interaction among elementary particles which causes atomic nuclei and elementary particles to decay.

White dwarf A small-radius, high-density star in its final stage of evolution which is on the verge of becoming extinct. *See also* Neutron star.

World of desire The lowest of the threefold world, so called because its inhabitants are ruled by desires.

X-rays Short-wavelength electro-magnetic radiation falling between the ultra-violet and gamma-ray portions of the electro-magnetic spectrum.

Yin-yang teachings In Chinese cosmology, it is thought that the universe was generated and its harmony is maintained through the interaction of the two forces of yin and yang. Yin is dark and negative, and yang is light and positive.

Yojana (Skt): a unit of measurement in ancient India, equal to the distance which the royal army was thought to march in a day. Approximations vary as widely as 9.6, 18 and 24 kilometres.

Zodiac The orbital circle of the Sun, called the ecliptic, divided into twelve parts or signs of the zodiac of thirty degrees each, represented by animals or other beings. This system of divining was developed by the Greeks based on the concept that there is a relationship between the macrocosm or the universe, and the microcosm or man.

Zōkyō, tsūgyō, bekkyō and engyō A classification of Shakyamuni's lifetime teachings according to their content. They are generally known as the four teachings of doctrine.

Materials used for references: *McGraw-Hill Dictionary of Scientific and Technical Terms,* Third Edition; *Concise Science Dictionary,* Oxford University Press; *The Penguin Dictionary of Physics; Longmans English Larousse; The Penguin Dictionary of Astronomy; A Dictionary of Life Science,* Pan Books; *The New Encylopaedia Britannica,* Fifteenth Edition; *A Cultural Dictionary of Japan,* The Japan Times, Ltd; *Kodansha Encyclopedia of Japan; A Dictionary of Buddhist Terms and Concepts,* Nichiren Shoshu International Centre.

Index

Numbers in brackets refer to footnotes.

action property, 39
air element, *see* wind ...
Akutagawa, Ryūnosuke, 298
ālaya-consciousness, 308-309
Alexander the Great, 326, 328
Andromeda Nebula, 59, 67, 68
anger, 46-7, 181-3
animality path, 46, 47, 173, 182
animals, death sense, 303-4
anthrophic principle, 21
Apollo missions, 125-6, 218
archaeo-astronomy discipline, 110
Aries-Taurus meteor swarm, 168-9
Aristotle, four elements, 171
Armstrong, Neil, 218
Article Nine, 281-2, 290-1(7)
astrology, 16
astronauts, 102-5, 217-18, 242-6
   extra-vehicular activity, 292
astronomy, 11-12, 86-7, 247-8
   ancient astronomers, 108-14
   comet events, 155-63
   cosmological concepts, 190
   discoveries, procedure, 126-7
   E.T. search projects, 88-94
   Mars canals myth, 299-300
   phenomena, 163-4
   photography, 130, 133-4, 267
   planet estimates, 59, 84, 87-90
   solar systems, 82-3, 226-31
   telescopes, 130-3, 228, 267-9
   Venus, 192-9
asura, meaning, 46-7
atomic energy, 269-73

attitudes, state of mind, 200-1
Auguries of Innocence, 223

Bāsho (artist), 58
beginningless time, 71-4
behaviour, lunar influence, 201-2
benevolence, 283
Bhagavat, 51
Big Bang theory, 11, 96, 318
birth/death, cosmic scale, 313-17
black holes, 41-3, 44
   satellite sighting, 251, 265(4)
   time distortions, 63-4
Blake, William, 223, 224
Bodhisattva state, 48
Bohr, Neils (1885-1962), 154-5
Bon festival, 220, 239(2)
bonnō, meaning, 283-4
Buddha
   death event, 148-9
   eye of, 136, 139-41
   illness/four elements, 170-1
   life span, 185
   names, 51, 84
   wisdom, meaning, 38
Buddhahood, 48, 310
Bun'ei era, significance, 163-4
buppō (Buddha's Law), 69
Bussard, Robert W, 67
butsumetsu, meaning, 189-90

Calder, Nigel, 42, 44, 79
calendars, 219-26
calligraphy interlude, 222-3

capital punishment, 146-50
Carnap, Rudolph (1891-1970), 155
Carr, Gerald P, 242-6, 294
causality, 143, 309
cause and effect, 68, 70, 103
   inevitability, 147-8
   scientific outlook, 311-12
Ceremony in the Air event, 14
Chandrasekhar, S, 11, 239
change
   seasonal rhythm, 240-2
   trigger effect, 231-5
   truth prejudices, 332-3
chanting, purpose, 264, 286, 288-89
chemistry of life, 75-8, 94, 213-16
Chih-i, see T'ien-t'ai
Chinese astronomy, 110-11
Chinese calendar, 225-6, 239(1)
Christianity, 23, 27, 143
chūtai, meaning, 260, 261
Cipango search, 249, 265(3)
civilizations, other planets, 59, 84, 87-91
Clark, William S. (1826-1886), 122
clock paradox, see Urashima ...
colour
   Earth from space, 245
   outer space, 54
   spectral types, 92, 100(5)
comet events, 127, 155-63, 168
compassion, need, 284
consciousnesses, 307-11
   six functions, 288, 289
   seventh, 308
   eighth, 308-9,
   ninth, 288, 289, 310
constellations, names, 112-13
continuance kalpa, 116, 153
cosmic dust, 76-7
cosmic religious sense, 23-4
cosmonauts, 102-5, 217
cosmos, 57-9, 116
Crab Nebula, 315
curvature of space, 61-2
Cyclops project (NASA), 88, 100(5)

Daibyakurenge, 98, 101(7)

daimoko chant, 264
Daishōnin, Nichiren, 14, 30
   astronomical phenomena, 163-4
   darkness, 185, 186
   death, 146, 148, 149, 206-9
   environment and life, 155
   European recognition, 249-51
   falling star account, 196-8
   five types of vision, 140, 141
   four elements, 170-1
   karma, 147
   life-moment truth, 69, 70
   myō and *hō*, 209, 210
   nam-myōhō-renge-kyō, 264
   ninth consciousness, 58-9
   prayer attitude, 26, 28
   proof principle, 134, 137
   stupidity/animality link, 182
   Tatsunokuchi events, 164-70
   Ten Worlds, explanation, 44
   worship choice, 27
danger years, 32, 50(1)
darkness, 184, 284
The Day After (film), 273-4
death, 142-80, 202-6
   Buddhist approach, 142-4, 206-9
   cosmic scale, 313-17
   equality state, 305
   execution scares, 204-5
   fear aspect, 188-90, 192, 306
   Japanese view, 186-92
   Kobayashi's experience, 19-20
   myō representation, 209-12
   religious consciousness, 303
   wildlife awareness, 303-4
decline kalpa, 116, 153
deductive reasoning, 15-17
Delacroix paintings, 223
delusion, 184, 310
desire, 45-6, 65, 283
   suffering, elimination, 180-6
   transformation, 286
   world of, 40
Devil of the Sixth Heaven, 184, 286
devotion attitude, 26, 181
dharma (vessel of the Law), 146
Dirac, Paul A.M., 258

disentigration kalpa, 116, 153
distance, 37, 48-9, 57-8
    human perception, 135-6
    nearest sun class star, 92
divine eye term, 136-8
Dōji *see* Sessen Doji
dōri, meaning, 102-3
Dostoyevsky, death scare, 205-6
Drake, Frank D, 80, 84, 89-90
Draper, John W. (1811-1882), 134

Eagle Peak assembly, 14
Earth
    age estimate, 151-2
    astronaut's view, 244-5
    beauty and colour, 51-3
    'cradle of human life', 104
    formation, 213-16, 226-7
    last stages, 150-3, 251-5
    life origins, 10-11, 77-80
    revitalization need, 169-70
    satellites, man-made, 267-8
    shape, early ideas, 329-31
    spaceship concept, 86-7, 320
earth element, 170
Eatherly, Claude, 280-1, 291(4,5)
Echi, falling star event, 195-9
Edo period, 40, 60
Egyptian pyramids, 327-8
eighty thousand teachings, 117
    kū concept, 260
Einstein, Albert, 15
    cosmic religious sense, 23-4
    energy equation, 271
    grand unified theory, 117, 259
    space-time concept, 61
elementary particles, 317
Emperor Daigo poem, 305
Encke's Comet event, 168
energy
    cosmic scale, 315-16
    nuclear, 269-73
enlightenment, 65, 140, 184
    earthly desires, 288
    good and evil inherence, 285-6
    myō indicator, 285
    ninth consciousness, 310

Shakyamuni's decision, 208
Enomoto, Takeaki (1826-1908), 122
environment, oneness, 155
eshō funi principle, 30, 184
E.T. movie, unreality, 55
E.T. *see* extraterrestrials
eternity
    Bhudda's life span, 185
    Buddhist view, 53, 65, 68-73
    life and death cycle, 150
    moment doctrine, 53, 69-71
Evershed, John (1864-1956), 248
evolution of life, 9-11, 214-16
existence, 39, 153
    Buddhist mission, 177-99
    illusions, 283, 284
    kū latency, 259
    three truths (properties), 260-3
expanding universe, 22, 83-4, 86-7
explorers, 326-8
external world, 200-1, 320-35
extra-sensory perception, 137
extraterrestrials, 55, 300-1
    Buddhist view, 35-8, 56
    existence question, 82-101
    Martian tales, 298-300
    Nacza drawings claim, 321
    probability, 59, 213-16, 245-6
    search projects, 88-94
eye types in Buddhism, 135-41

Fa-hsien journeys, 327
faith, Mystic Law, 72, 191
fear of death, 188-90, 192, 306
February, significance, 224-5
festivals, New Year's Day, 219-24
fire element, 170-2
five impurities, 183-4
five types of vision, 135-41
flags, significance, 127-9
form, world of, 140
formation kalpa, 116, 153
formlessness, world of, 140
four continents realm, 84
four elements, 170-2, 228
four kalpas, 116, 140, 153, 314
four lower worlds, 45-7

four noble worlds, 47-8
four sufferings, 19, 140, 170, 208
Fugen Sutra teaching, 98, 121
Fukuzawa, Yukichi, educator, 295
Fundamental Law, 25-50, 208
Fushimi, Kōji, astronomer, 11

Gagarin, Yuri, 217
Galaxy death, 255-6
Galileo, 85-6, 331-5
Gamow's theory, 20-2
Gauss theorem on shape, 24
Genghis Khan, 327, 328, 329
geocentric theory, 331-5
Goddard, Robert H., 218, 219
God(s), 27-8
Gohonzon, 27-8, 120-1
gongyo ceremony, 79, 120-1, 175
good and evil, 184, 285-6
grand unified theory, 116-19, 259
gravitation, 41-3, 61, 63-4, 80
greed, 181-3
Gregorian Calendar, 224, 241
Guimet Museum collection, 250

Hachiman Shrine, 27-8
haiku, 58
Halley's Comet events, 155-60
happiness, 102, 297
harmony of life, 129-30, 240-2
Hawin, R., 48
Hawking, Stephen W., 43, 96, 100-1(6)
Hawkins, Gerald S., 110
Hayakawa, Sachio, astronomer, 11
Hayashi, Chūshirō, 11, 213, 227
heaven, 47
Heisenberg, Werner, 154-5
hell, 43-4, 50(4), 173-4, 305
Herschel, Sir Frederick W., 86
Hidaka, Yoshiki, peace, 276-7
Hinayana, Buddhism, 118, 285
Hirose, Hideo, 168, 197-8
Hiroshima bomb, 272, 275, 279-80
hō, 79, 209-10, 285
Hoba West Meteorite, 124
Hōben chapter, 38, 183-4

Hokke sect, 183-4, 286
Holland, telescope invention, 324
honchi nanshi, of Buddha, 97
Honzon, meaning, 26-7
hope, importance to life, 181
Hoyle, Fred, 194, 195, 297
Hsuan-tsang, 17-year journey, 327
Hubble, Edwin, 86
human beings, 320
  black hole event, 42, 63-4
  body, 30-1, 45, 301
  Buddha nature, 312
  danger years, 32, 50(1)
  devotion to life, 181
  evil side, 185, 286
  four kalpas cycle, 153
  Fundamental Law, 25-50
  life span (age), 185-6
  lunar influence, 201-2
  ninth consciousness, 58-9
  origin theories, 10-11
  religious consciousness, 303
  senses, 95-6, 307
  transience, 35-6, 64, 65
humility, essential nature, 40
hunger, 45-7, 173

ichinen principle, 13-14, 28, 30, 181
ichinen sanzen, 13, 14, 19
  good and evil, 184, 285
  prayer, 257
  universe explanation, 84
ignorance and stupidity, 182
illness, Buddhist view, 170-2
illusion, 183, 283-4
immortality elixir, 60, 81(4)
impermanence doctrine, 65
impurities, 183-4
indeterminacy principle, 154-5
inductive reasoning, 15-17
infra-red telescopes, 132, 228, 267, 268-9
inherent truth principles, 103
inner world of mind, 9-24
inspiration, suddenness, 232-4
intercalary month, 224-5
interstellar travel, 67-8, 72-4

intuition, 119, 137
IRAS-Araki-Alcock Comet, 127, 160

Japan, 240-2
  ancient astronomy, 108, 110
  Article Nine, 281-2, 290-1(7)
  astronaut recruitment, 217
  astronomers, 268-9, 290(1, 2)
  calendars, 225-6, 239(1), 241
  Cipango name, 249, 265(3)
  Halley's Comet events, 159-60
  life and death view, 186-92
  Mongol invasions, 146, 175-6(2)
  needles ceremony, 188
  New Year's Day, 219, 239(1,2), 241-2
  peace movements, 275-82
  space programme, 321-2
  suicides increase, 180-1
Jarrow, Robert, 216
Judaism, prayer, 27
Juryō chapter. 38-9, 87-8, 185
  life and death cycle, 150
justice, 283

kalpas, 116, 140, 153, 314
kalpa's end, 14, 24(1), 314
Kanji chapter, 166-7
Kant, Immanuel, 12, 13, 20
  Japan essay, 249-50
karma 53, 143, 147, 281
  cosmic manifestations, 309
  eighth consciousness, 308-9
  life span effects 185-6
  negative and rebirth, 173
kasei (Mars), 296
kayomi (koyomi), 225, 239(4)
keiwaku (Mars), 296
Kepler, Johann, 130, 138
kimyō (to devote one's life), 26
Kissinger, Henry, 275
Kobayashi, Hideo, 19-20, 207
Kokin wakashū, 240, 265(1)
Kraus, John, 89, 95
kū concept, 39, 115-16
  Buddhist teaching, 256-63, 309

human life cycle, 153
kuon ganjo (Mystic Law), 71-2

Laplace, Pierre Simon, 213
The Last Day, 252
latency, 143, 153, 175(1)
  kū aspect, 259, 260
Latter Day of the Law, 71, 72
  fundamental seed revealed, 85
  Law slander, 166
  life and death problem, 208-9
  three poisons taint, 183, 184
Law of the universe, 13, 142-6
  eye of, 136, 139-41
  impermanence, 65
  nam-myōhō-renge-kyō, 264
  three properties, 39
Learning life-state, 48
Lexell's Comet, 161
life, 143, 188
  Buddhist approach, 35, 53, 55-6
  emergence theories, 9-11, 214-16
  eternal state, 69-70, 209
  extraterrestrial, 36-7, 55, 82-101, 213-16
  hō representation, 209-10
  ichinen usage, 14
  impurity state, 184
  kū, 259-63
  moment of, 16, 69, 70
  oneness principle, 155
  origins mystery, 74-5, 77
  peace/happiness, 29-35
  potential, 240-65
  protein synthesis, 77-8
  pulsation principle, 119-21
  regeneration, 173-5
  reverence teaching, 56-9
  self continuation, 261-2
  span, 185-6
  subjective approach, 149
  true nature, 14, 34, 186-92
life and death, 142-4, 200-16
  Buddhist teaching, 208-9
  eternal cycle principle, 150
  three truths, 260-3
light-year measurement, 37

Logunov, Anatoliy A., 10, 37
longevity, recognition for, 186-7
Lotus Sutra, 13-14, 38-9, 50(2)
 five impurities, 183-4
 juryō/external life, 185
 Kant's interest, 249-50
 life and death cycle, 150
 major world systems, 87-8
 Muryōgi-kyō and self, 34-5
 predictions, 166-7
 propagating Buddhism, 98
 Tatsunokuchi events, 167
 three poisons, 183
 threefold world truth, 149
 truth, meanings of, 52
Loux, Françoise, 45
Lowell, Percival, 248, 299
Lucien, moon travel story, 249

macrocosm viewpoint, 30-2, 38, 325
 revitalization need, 169-70
Maejima, Shinji (1903-83), 330-1
Magellan, world voyage, 323
Magellanic Cloud black hole, 251
Mahayana Buddhism, 13, 29
 illusions teaching, 285
 kū concept, 153
 Mystic Law approach, 84
 proof, importance, 134-5
Malraux, André, meeting, 306-7
mandala, 26, 264
manō-consciousness, 308
Man'yōshū, 60, 81(3)
Mars, observations, 295-302
Matheson, Richard, 252
matter potential, 256-63
Maudgalyāyana story, 45-6, 137
Maxwell, Robert, 293
Meiji era, new calendar, 241
Meiji Restoration (1868), 40, 122
Mesopotamian astronomy, 112
meteorites, 122-7, 168
Miao-lo, *myō* term, 211
microcosm viewpoint, 30-2, 38, 325
 revitalization need, 169
microscope invention, 324, 325
Middle Way (chu), 39-40, 115, 260

Milinder, Buddhism debate, 326
Milky Way, 59, 80
mind, 13-14, 18
 expansion, 119-20
 sickness, 171-2
mission of Buddhists, 177-99, 212
Mitsuzawa Stadium speech, 279
mo legend, 175, 176(5)
moment principle, 53, 69-74, 207
Monod, Jacques (1910-1976), 97-8
Montgomery, John D., peace, 287
Moon, 125-6, 150-3
 human influence, 201-2
 mapping changes, 130-1
 rocks loan, 294
moral law within, 9-13
Mother (poem), 203-4
Mount P'eng-lai, 60, 81(4)
Mount Sumeru theory, 40-1, 118
murder, 180, 305
Muryōgi Sutra, 34, 96
Muslim prayer, 27
myō, 79, 209-12, 285
Mystic Law, 71-2, 84-6
 Buddha nature awakening, 312
 eye of the Buddha, 139-41
 Juryō chapter, 38
 myō and *hō*, 209, 211
 nam-myōhō-renge-kyō, 288-9
 perfection principle, 107
 purification, senses, 95-6
 suicide pointlessness, 173
 tranquil awareness, 140-1

Nacza drawings, Peru, 321
Nagasaki bomb, 273, 275
nam-myōhō-renge-kyō, 85, 96, 97
 chant practice, 286, 288-9
 deceased persons, 174-5
 enlightened wisdom, 286
 essence of Buddhism, 264
 ninth consciousness state, 289
 three truths expression, 263
 ultimate principle, 238
namu(nam), meaning, 26, 28
NASA Moon rocks loan, 294
Nature, 238

nebular hypothesis, 213-15
Needham, Joseph, 326
needles ceremony, 188, 199(1)
Nembutsu sect, 189
neutron stars, 236-7
New Year's Day, 219-24, 239(1)
New York Times error, 218-19
Newton, Sir Isaac, 13, 61-2, 138
   boils his watch, 107
   universal gravitation, 270-1
Nichiji (b.1250), 330-1
Nichiren Shōshū Buddhism, 27
Nihon no okikata ..., 281, 290(6)
nine consciousnesses, 307-11
Ninnō Sutra, 84
ninth consciousness, 288, 289, 310
nirvana, 148, 150, 187
Nirvana Sutra, 234-5
Nobunaga, Oda, 268-9, 290(1)
non-substantiality (*kū*), 39, 153
nuclear fusion, 272, 290(3)
nuclear threat, 66, 266-91

objectivity, 18
Oda, Minoru, astronomer, 11
oneness principle, 13-14, 30, 45
   eternity concept, 69-71, 73
   five impurities, 184
   life/environment, 155
   three truths (properties), 263
Oppenheimer, J. Robert, 66
organ transplants, 64-5
Orion Nebula discovery, 82-3
Oshima, Tairō, 90
outer world of universe, 9-24
OXMA project, 88, 100(4)

Page, Glenn D., peace, 287
peace, 66, 102
   Buddhist philosophy, 266-91
   leadership need, 275-8, 293
   philosophy of, 56, 107-8, 278
   state of mind ..., 283
Penzias, Arno Allan, 22, 24(2)
perfection principle, 107
perversity, 46
photography, space, 130, 133-4

Pillow Book, 247, 265(2)
Pioner 10 probe, 93-4, 119, 134
planets, 59, 213-15, 226-7
poet laureate honour, 58, 81(1)
poisons, 181-3
Ponnamperuma, Cyril, 213, 214
potential (*kū*), 258, 259-63
power, diabolical nature, 205-6
prayer, 25-9, 257
   deceased persons, 173-5
   gongyo, purpose, 120-1
proof principle, 134-5
propagation, Buddhism, 84-6, 98-9
properties of existence, 39, 260
protein synthesis, 77-8
proton decay, 317-19
Pugwash Conference, 280, 293
pulsars, 315-6
pulsation principle, 119-21
purpose, sense of, 212

quantum mechanics, 236-7, 238-9

Rabbit and Moon tale, 87, 100(2)
radio-telescopes, use, 92-3
rapture (joy), 47
Reagan, Pres. Ronald, 322-3
reality, 65, 263
   death, 202-6
   kū, state, 259
   ninth consciousness, 58-9
realization life-state, 48
reason, discussions, 15-17, 102-3
rebirth teaching, 173-5
red giant stars, 253-4
reform principle, 64-6
Reichenbach, Hans (1891-1953), 155
religions
   ancient cultures, 108, 110
   burial, early societies, 303
   life and death question, 19-20
   prayer, 27
   science need, 29, 96-8, 216, 293
   space age, 328-35
   validation need, 134-5, 137
Ride, Sally, 103-5

Rissho antoku ron, 163-4
Russell-Einstein paper, 280, 293

Sagan, Carl, 12, 74, 82
    television series, 9, 19
Saitō, Mokichi, 203, 204
Sakurai, Kunimoto, 119
Sandage, Allan R., astronomer, 12
satellites
    black hole sighting, 251, 265(4)
    Earth orbiting, 267-8
    infra-red, 228, 239(6)
    Mars landing, 301
Schaeberle, J.M. (1853-1924), 248
Schiaparelli, G.V., 299
scholars, death ideas, 202-6
science, 9, 31-2, 320
    Buddhism search, 154-5
    diabolical side, 66-7
    discovery progress, 323-8
    inductive reasoning, 15-17
    life and death problems, 16-17
    limitations, 29, 159
    proof principle, 134-5
    religion need, 15, 293-4
    responsibilities, 107-8, 319
seasons, change rhythm, 240-2
self
    Buddhist teaching, 34-5, 307-11
    existing essence, 261-2
    kū state, 257
    perfection aim, 264, 289
senses, 95-6, 307
Sessen Dōji legend, 234-5
SETI project, 55, 83, 86-7, 100(4)
shakubuku, use 98-9
Shakyamuni Buddha
    eye of the Buddha, 139-40
    Lotus Sutra teachings, 13-14
    renouncement decision, 208
    Sessen Dōji legend, 234-5
    Tripitaka teaching, 85
Shibuya planetarium, 114
shimazaki, Toson (1872-1943), 203
shining object account, 164-70
Shining Princess legend, 87, 100(1)
Shōnin, Nichiko, teachings, 84-5

Shōzan, Sakuma, 40
Shrimala Sutra, teaching, 98
six consciousnesses, 288, 289
six paths, 47-8
Skylab space project, 242-6
Sōka Gakkai, 82, 85, 101(7), 255
    anti-nuclear exhibitions, 275-6
    peace movement, 278, 287, 290
Sōka Gakkai Internat. (SGI), 8, 128
Sōkanmon sho, 284, 311, 312
solar systems, 82-3, 226-31
South American visit, 322, 335(1)
space, 105-8
    black holes, 41-3, 44
    composition, 227
    curvature effect, 62
    distance measures, 37, 48-9
    environment conditions, 53-5
    human interest, 105-6
    kū source of matter, 256-8
    organic molecules, 75, 76-7
    solar systems, 82-3, 226-31
    warping/distortion, 61-2
space age religion, 329-36
space travel, 32-3, 49
    human achievements, 102-5
    Japanese plans, 321-2
    Skylab project, 242-6
    solar system, 53-4
    space age dawn, 217-19
    twins paradox, 67-8, 72-4
space-time, 61, 72
spaceship travel, 67-8, 72-4
Spinoza, Baruch, 283, 291(8)
spiritual awakening, 232-5
spring season, 240-2
Ssu-ma Ch'ien, 111, 296
stars, 59
    ancient records of, 110-13
    black hole stage, 251-2
    brightness measurement, 156-7
    classification, 91-2, 100(5)
    daytime appearances, 195-6
    death, 235-7, 313-17
    distance measurement, 135-6
    formation, 229-31, 256-9
    life of, 217-39

life spans, 150-1
red giant stage, 253-4
shape, 24
Stonehenge, 110
Strömgren, Bengt, G.D., 11
stupidity, 181-3
subjective time, 68-9
subjectivity, 18, 200-1
suffering, 53, 63
  four aspects, 19, 140, 170, 208
suicide
  Buddhist view, 144-6, 173
  increase trend, 179-82, 305-6
  Nembutsu sect, 188
Sun
  colour grade 92, 100(5)
  eclipse (1983), 52, 105
  formation, 226-7
  last stages, 152, 251-2
  life energy source, 79, 81(6)
  life span, 150-3
  subjective size, 200-1
  worship cultures, 108
supernovae, 256, 316
sutras, 13-14, 24
  eighty thou' teachings, 117
  reason, explanation, 103
  scale of universe, 84

Ta yen li (Calendar), 225-6
Takamatsuzuka tomb, 110
Tale of Genji, 146-7, 176(3)
tanka (poem), of spring, 240
Tathāgata, 51, 69, 140, 149
Tatsunokuchi events, 164-70, 176(4)
Teika, Fugiwara no, diary, 161-2
telescopes, 130-3, 324-5
  infra-red, 228, 267-9
Teller, Edward, 66
temperature, space, 55, 316-17
Temple of Karnak, 110
temporary existence (ke), 39
Ten Worlds doctrine, 14, 35, 41-5
  life migration, 150
  self, existence, 309
tengen (divine eye), 136-8
Tenma discovery, 251, 265(4)

Tereshkova, Valentina V.N., 103-5
The Universe, (poem), 68-9
thought, 284, 308
three evil paths, 45-7, 173-4
three poisons of life, 181-3
three truths, 39, 260-1
threefold world, 140
T'ien-t'ai, 14, 19, 98, 120
time, 68-74
  alteration in space, 61-2
  black holes, 63
  calenders, 219-26, 240-2
  twins paradox, 49, 59-60
Toda, Jōsei, 84
  Galaxy death, 255
  human transience, 35-6
  Ikeda's teacher, 15, 21
  nuclear weapons, 279-82
  religion and science, 216
  three poisons principle, 181
Tokugawa Ieyasu, 268, 290(2)
Toynbee, Arnold
  Daishonin writings, 250-1
  Eastern civilizations, 326
  higher religions, 29
  live for the moment, 207
  Lotus Sutra, teachings, 13
Toyoda, Toshiyuki, 283
tranquil awareness, 140-1
transformation, suddenness, 231-5
Tripitaka teachings, 85
truth, 39, 102-3
  Buddhist terms, 52
  illusion element, 283
  moment of, 233-5
  reactions to, 332-3
  subjective approach, 149
  three (properties), 39, 260-1
Turner, H.J. (1861-1930), 248
twins paradox, *see* Urashima ...

Uchinoura prayer story, 25-6
ultimate truth (Law), 17
UNESCO charter on war, 311
unhappiness, 183, 297
unified field theory, 116-19, 259
United Nations, 127-9, 275

universal values, need, 320
universe, 29, 321-36
   anthrophic principle, 21
   birth and death, 293-320
   Buddhist views, 17-19, 35, 68-74, 83-5
   expanding theory, 20-2, 86-7
   four elements composition, 172
   Galaxy death, 255-6
   heartstrings concept, 105-8
   Law/ultimate truth, 17
   life hypothesis, 213-16
   myōhō reality, 237-8
   mysterious nature, 78-80, 115
   origin theories, 10-11, 86-7
   other worlds, 56, 59, 80, 84, 87-90
   scale comparisons, 57-9
   solar system, origins, 226-31
Urabon, ceremony, 49, 50(5), 138
Urashima effect, 49, 59-60, 61
   space travel, 67-8, 72-4
Urashima Tarō tale, 60-1, 81(2)
Ushio magazine, 8

Vasco da Gama, India route, 324-5
Vasubandhu, Buddhist scholar, 118
Venus, mysteries of, 192-9
vijnāna, 308
vision, Buddhism approach, 135-9

war, 36, 245
   murder standpoint, 180
   nuclear threat, 273-83
   UNESCO charter, 312
water element, 170-2, 228, 239(5)
Watson, Lyall, 201
Wei chih, 225, 239(3)
Weisskopf, Victor F., 66
Wells, H.G., 300
Wheeler, John A., astronomer, 11
white dwarfs, 235
Whitehead, Alfred (1861-1947), 155
Wilson, Robert Woodrow, 22, 24
wind, solar, 54
wind element, 170-2
wisdom, 39, 136, 138-9
women cosmonauts, 102-5
world unity need, 106
worship, 26, 108
   Middle Way object, 40
   nam-myōhō-renge-kyō, 264

X-ray telescopes, 77, 81(5), 268

young people
   education need, 247-8, 295-6
   violence, 99-100
Younghusband, Sir Francis E., 95
yuchou (universe), 114, 118
Yukawa, Hideki, 11, 21

RL E102RRR
IKEDA